Forests

Environmental Issues, Global Perspectives

Forests

James Fargo Balliett

Routledge
Taylor & Francis Group

LONDON AND NEW YORK

First published 2010 by M.E. Sharpe

Published 2015 by Routledge
2 Park Square, Milton Park, Abingdon, Oxon OX14 4RN
711 Third Avenue, New York, NY 10017, USA

Routledge is an imprint of the Taylor & Francis Group, an informa business

Notices
No responsibility is assumed by the publisher for any injury and/or damage to persons
or property as a matter of products liability, negligence or otherwise, or from any use
of operation of any methods, products, instructions or ideas contained in the material herein.

Practitioners and researchers must always rely on their own experience and knowledge in
evaluating and using any information, methods, compounds, or experiments described herein.
In using such information or methods they should be mindful of their own safety and the
safety of others, including parties for whom they have a professional responsibility.

Product or corporate names may be trademarks or registered trademarks, and are used only
for identification and explanation without intent to infringe.

Library of Congress Cataloging-in-Publication Data

Balliett, James Fargo.
Forests: environmental issues, global perspectives / James Fargo Balliett.
    p. cm.
Includes bibliographical references and index.
ISBN 978-0-7656-8227-7 (hardcover: alk. paper)
1. Forests and forestry—Environmental aspects. I. Title.

SD387.E58B35 2010
333.75—dc22   2010012120

Figures on pages 14, 26, and 46 by FoxBytes.

ISBN 13: 9780765682277 (hbk)

Contents

Over the past 150 years, the planet Earth has undergone considerable environmental change, mainly as a result of the increasing number of people living on it. Unprecedented population growth has led to extensive development and natural resource consumption. A population that numbered 978 million people worldwide in 1800 reached 6.7 billion in 2009 and, according to the United Nations, may well exceed 8 billion by 2028.

This sixfold growth in population has brought about both positive and negative outcomes. Developments in medicine, various natural and social sciences, and advanced technology have resulted in widespread societal improvements. One measure of this success, average life expectancy, has climbed substantially. In the United States, life expectancy was 39.4 years in 1850; by 2009, it had grown to 77.9 years.

Another major change for global nations and cultures has been the accessibility and sharing of information. Though once isolated by oceans and geography, few communities remain untouched by technological innovation. The construction of jumbo jet planes, the development of advanced satellites and computers, and the power of the Internet have made travel and information readily available to an unprecedented number of people.

Technological advances have allowed residents of just about anywhere on the planet to share events and information almost instantly. For example, images of an expedition on the summit of Mount Everest or a scientific investigation in the middle of the Atlantic Ocean can be posted online via satellite and viewed by billions of citizens worldwide. In addition, cumulative and individual environmental impacts now can be assessed faster and more comprehensively than in years past.

During the same period, however, regulated and unregulated residential and business development that consumes natural resources has had profound impacts on the global environment. Such impacts are evident in the clear-cutting or burning of forests, whether this deforestation is for fuel or building materials, or simply to clear the land for other activities; the drainage of wetlands to divert freshwater, expand agriculture, or provide more building upland; the overfishing of oceans to meet an ever-growing appetite for seafood; the stripping of mountaintops for fuel, metals, or minerals; and the pollution of freshwater sources by the waste output of industrial and residential communities. In fact, pollution has spread across land, air, and water biomes in ever-increasing concentrations, causing considerable damage, especially to fragile ecosystems.

The Emergence of a Global Perspective

Environmental Issues, Global Perspectives provides a fresh look at critical environmental issues from an international viewpoint. The series consists of five individual volumes: *Wetlands, Forests, Mountains, Oceans,* and *Freshwater.* Relying on the latest accepted principles of science—the acquisition of knowledge based on reasoning, experience, and experiments—each volume presents information and analysis in a clear, objective manner. The overarching goal of the series is to explore how human population growth and behavior have changed the world's natural areas, especially in negative ways, and how modern society has responded to the challenges these changes present—often through increased educational efforts, better conservation, and management of the environment.

Each book is divided into three parts. The first part provides background information on the biome being discussed: how such ecosystems are formed, the relative size and locations of such areas, key animal and plant species that tend to live in such environments, and how the health of each biome affects our planet's environment as a whole. All five titles present the most recent scientific data on the topic and also examine how humans have relied on each biome for survival and stability, including food, water, fuel, and economic growth.

The second part of each book contains in-depth chapters examining seven different geographically diverse locations. An overview of each area details its unique features, including geology, weather conditions, and endemic species. The text also examines the health of the natural environment and discusses the local human population. Short- and long-term environmental impacts are assessed, and regional and international efforts to address interrelated social, economic, and environmental issues are presented in detail.

The third part of each book studies how the cumulative levels of pollution and aggressive resource consumption affect each biome on a global scale. It provides readers with examples of local and regional impacts—filled-in wetlands, decimated forests, overdeveloped mountainsides, empty fishing grounds, and polluted freshwater—as well as responses to these problems. Although each book's conclusion is different, scenarios are highlighted that present collective efforts to address environmental issues. Sometimes, these unique efforts have resulted in a balance between resource conservation and consumption.

The *Environmental Issues, Global Perspectives* series reveals that—despite the distance of geography in each title's case studies—a common set of human-induced ecological pressures and challenges turns up repeatedly. In some areas, evidence of improved resource management or reduced environmental impacts is positive, with local or national cooperation and the application of new technology providing measurable results. In other areas, however, weak laws or unenforced regulations have allowed environmental damage to continue unchecked: Brazen frontiersmen continue to log remote rain forests, massive fishing trawlers still use mile-long nets to fill their floating freezers in the open ocean, and communities, businesses, vehicles, factories, and power plants continue to pollute the air, land, and water resources. Such existing problems and emerging issues, such as global climate change, threaten not just specific animal and plant communities, but also the health and well-being of the very world we live in.

Wetlands

Wetlands encompass a diversity of habitats that rely on the presence of water to survive. Over the last two centuries, these hard-to-reach areas have been viewed with disdain or eliminated by a public that saw them only as dangerous and worthless lowlands. The *Wetlands* title in this series tracks changing perceptions of one of the world's richest and biologically productive biomes and efforts that have been undertaken to protect many areas. With upland and coastal development resulting in the loss of more than half of the world's wetlands, significant efforts now are under way to protect the 5 million square miles (13 million square kilometers) of wetlands that remain.

In *Wetlands*, three noteworthy examples demonstrate the resilience of wetland plants and animals and their ability to rebound from human-induced pressures: In Central Asia, the Aral Sea and its adjacent wetlands show promising regrowth, in part because of massive hydrology projects being implemented to undo years of damage to the area. The Everglades wetlands complex, spanning the lower third of Florida, is slowly reviving as restorative conservation measures are implemented. And Lake Poyang in southeastern China has experienced increased ecological health as a result of better resource management and community education programs focused on the vital role that wetlands play in a healthy environment.

Forests

Forests are considered the lungs of the planet, as they consume and store carbon dioxide and produce oxygen. These biomes, defined as ecological communities dominated by long-lived woody vegetation, historically have provided an economic foundation for growing nations, supplying food for both local and distant markets, wood for buildings, firewood for fuel, and land for expanding cities and farms. For centuries, industrial nations such as Great Britain, Italy, and the United States have relied on large tracts of forestland for economic prosperity.

The research presented in the *Forests* title of this series reveals that population pressures are causing considerable environmental distress in even the most remote forest areas. Case studies provide an assessment of illegal logging deep in South America's Amazon Rain Forest, a region closely tied to food and product demands thousands of miles away; an examination of the effect of increased hunting in Central Africa's Congo forest, which threatens wildlife, especially mammal species with slower reproductive cycles; and a profile of encroachment on old-growth tropical forests on the Southern Pacific island of Borneo, which today is better managed, thanks to the collective planning and conservation efforts of the governments of Brunei, Indonesia, and Malaysia.

Mountains

Always awe-inspiring, mountainous areas contain hundreds of millions of years of history, stretching back to the earliest continental landforms. Mountains are characterized by their distinctive geological, ecological, and biological conditions. Often, they are so large that they create their own weather patterns. They also store nearly one-third of the world's freshwater—in the form of ice and snow—on their slopes. Despite their daunting size and often formidable climates, mountains are affected by growing local populations, as well as by distant influences, such as air pollution and global climate change.

The case studies in the *Mountains* book consider how global warming in East Africa is harming Mount Kenya's regional population, which relies on mountain runoff to irrigate farms for subsistence crops; examine the fragile ecology of the South Island mountains in New Zealand's Southern Alps and consider how development threatens the region's endemic plant and animal species; and discuss the impact of mountain use over time in New Hampshire's White Mountains, where stricter management efforts have been used to limit the growing footprint of millions of annual visitors and alpine trekkers.

Oceans

Covering 71 percent of the planet, these saline bodies of water likely provided the unique conditions necessary for the building blocks of life to form billions of years ago. Today, our oceans continue to support life in important ways: by providing the largest global source of protein in the form of fish populations, by creating and influencing weather systems, and by absorbing waste streams, such as airborne carbon.

Oceans have an almost magnetic draw—almost half of the world's population lives within a few hours of an ocean. Although oceans are vast in size, exceeding 328 million cubic miles (1.37 billion cubic kilometers), they have been influenced by and have influenced humans in numerous ways.

The case studies in the *Oceans* title of this series focus on the most remote locations along the Mid-Atlantic Ridge, where new ocean floor is being formed 20,000 feet (6,100 meters) underwater; the Maldives, a string of islands in the Indian Ocean, where increasing sea levels may force residents to abandon some communities by 2020; and the North Sea at the edge of the Arctic Ocean, where fishing stocks have been dangerously depleted as a result of multiple nations' unrelenting removal of the smallest and largest species.

Freshwater

Freshwater is our planet's most precious resource, and it also is the least conserved. Freshwater makes up only 3 percent of the total water on the planet, and yet the majority (1.9 percent) is held in a frozen state in glaciers, icebergs, and polar ice fields. This leaves only 1.1 percent of the total volume of water on the planet as freshwater available in liquid form.

The final book in this series, *Freshwater*, tracks the complex history of the steady growth of humankind's water consumption, which today reaches some 3.57 quadrillion gallons (13.5 quadrillion liters) per year. Along with a larger population has come the need for more drinking water, larger farms requiring greater volumes of water for extensive irrigation, and more freshwater to support business and industry. At the same time, such developments have led to lowered water supplies and increased water pollution.

The case studies in *Freshwater* look at massive water systems such as that of New York City and the efforts required to transport this freshwater and protect these resources; examine how growth has affected freshwater quality in the ecologically unique and geographically isolated Lake Baikal region of eastern Russia; and study the success story of the privatized freshwater system in Chile and consider how that country's water sources are threatened by climate change.

Acknowledgments

I owe the greatest debt to my wonderful Mom and Dad, Nancy and Whitney, who led me to the natural world as a child. Thank you for encouraging curiosity and creativity, and for teaching me to be strong in the midst of a storm. I also could not have gotten this far without steady support, expert advice, and humorous optimism from my siblings: Blue, Julie, Will, and Whit.

I greatly appreciate the input of Dr. Arri Eisen, Director of the Program in Science and Society at Emory University, at key stages of this project. My sincere thanks also go to the superb team at M.E. Sharpe, including Donna Sanzone, Cathy Prisco, and Laura Brengelman, as well as Gina Misiroglu, Jennifer Acker, Deborah Ring, Patrice Sheridan, and Leslee Anderson. Any title that explores science and the environment faces daunting hurdles of ever-changing data and a need for the highest accuracy. This series benefited greatly from their precise work and steady guidance.

Finally, *Environmental Issues, Global Perspectives* would not have been possible without the efforts of the many scientists, researchers, policy experts, regulators, conservationists, and writers with a vested interest in the environment.

The last few decades of the twentieth century brought a significant change in awareness and attitudes toward the health of this planet. Scholars and laypeople alike shifted their view of the environment from something simply to be consumed and conquered now to a viewpoint of it as a significant asset because of its capacities for such measurable benefits as flood control, water filtration, oxygen creation, pollution storage and processing, and biodiversity support, as well as other positive features.

Knowledge of Earth's finiteness and vulnerability has resulted in substantially better stewardship. My thanks go to those people who, through their visions and hard work, have taught the next generation that fundamental science is essential and that humankind's collective health is inextricably tied to the global environment.

James Fargo Balliett
Cape Cod, Massachusetts

INTRODUCTION
TO FORESTS

1 | The Forests Around Us

A forest is not simply a stand of trees. It is a living and breathing ecological community. Towering maples, oaks, and spruces dominate this landscape, but they are a minority when it comes to the total number of plant and animal species present.

An average forest may host anywhere from 10,000 total species in a temperate climate to several hundred thousand species in a tropical one. These species have adapted to thrive in unique microclimates within the forest, ranging from open woodlands to dense treetops.

The forest floor, in particular, thrives with life. It supports a plethora of species that bound, crawl, dig, fly, hop, roll, slither, and walk. One square acre (0.16 square hectare) of South American rain forest has been estimated to support more than 165,000 insect species.

Forests Defined

A forest is an ecosystem anchored by long-living woody vegetation. The number and density of species in a forest depend on the climate, topography, soil conditions, and availability of freshwater. Forests are found in both warm and cold locations, and they range in size from less than 1 acre (0.4 hectare) to 50 million acres (20 million hectares). Tree species live for very long periods. The oldest known individual specimen was a pine tree in eastern Nevada, estimated to be 4,862 years old when it was cut down in 1964 for research purposes. Trees also grow in groups, called clonal masses, due to their ability to generate genetic

Scientists study trees in the Tongass National Forest in southeast Alaska. As a result of logging since the 1950s, only 30 percent of the old-growth forest remains. Covering 26,250 square miles (67,988 square kilometers), the Tongass is the largest publicly owned forest in the United States. *(Melissa Farlow/National Geographic/ Getty Images)*

copies over time. The oldest known clonal root system is in Dalarna, Sweden, and it is approximately 9,950 years old.

A forest provides a set of measurable benefits that are known as ecological services. One of the greatest assets of a forest is the fact that it is able to take in large amounts of carbon dioxide and convert it to oxygen and energy in a process known as photosynthesis. A forest also cycles plant nutrients from the soil into living material and back again. Roots absorb nitrogen and eventually deliver it to leaves, promoting new growth. Once the leaves fall to the ground and decompose, the nitrogen is again processed into the soil.

A forest holds water and protects watersheds, thereby providing a stable environment for a diversity of species. Forests even limit weather events such as droughts. This is performed by individual trees that hold water over time, as well as by the network of closely growing species of trees and other plant species.

Forest ecosystems are always changing in a process known as succession. A tree species is not static; it will emerge, grow, and eventually die to be replaced by another tree. Succession can occur over a number of years or be triggered by a disturbance such as a flood or a fire. This level of change allows a forest to thrive and evolve.

Twenty thousand years ago, forests covered half of all land on Earth. This habitat has been severely reduced due to human-generated deforestation and development, especially over the last 250 years. Today, scientists estimate that forests make up approximately 28 percent of all land, covering 9 percent of the planet.

A total of 13.8 million square miles (35.7 million square kilometers) of forest is broken down into six different types of ecosystems, including boreal, temperate deciduous, temperate evergreen, tropical rain, tropical seasonal, and savanna. Forests and their associated wetlands host a documented 1.75 million species, an estimated 2 percent of all living creatures.

Up to 55 percent of all forests is located in temperate zones, with Russia and North America supporting up to two-thirds of the nearly 7.6 million square miles (19.7 million square kilometers) of temperate forests. The remaining 45 percent is found in tropical areas, with Africa, Central America, and South America holding nearly two-thirds of these 6.2 million square miles (16.1 million square kilometers) of forest.

Origins of the Forest

Although the Earth is approximately 4.6 billion years old, precursor forest species did not appear until 420 million years ago during the Silurian Period. Scientists believe that the first forests dominated by trees had not evolved until 225 million years ago.

The first life on the planet emerged roughly 3 billion years ago in the form of marine algae and unicellular floating organisms. The first plants on land included

PERCENTAGE OF FORESTED LAND, BY REGION

Region	Percentage Forested
Africa	24
Asia	19
Central America	26
Europe	29
North America	38
Pacific Islands	11
Russia	34
South America	47

Source: United Nations Forum on Forests, 2007.

the family of Rhyniophytes, which were composed of simple, upward-growing shoots, ending in spore capsules. These vascular species relied on thin root hairs rather than thick roots to absorb nutrients; this might have been a sign of their evolution from an aquatic environment, where hair-like roots are common. The plants relied on distribution of their spores to reproduce and only thrived in a warm, moist environment. Most species of plants did not exceed 3 feet (1 meter) in height during this time.

More than 345 million years ago, a second major evolution in plant forms began. Several families of plants developed larger leaf systems, thicker stems, and substantial roots. These included ferns (from the classes *Psilotopsida, Equisetopsida, and Polypodiopsida*) and horsetails (from the class *Equisetaceae*). Many of these species experienced accelerated growth; for example, the tree-sized club moss grew up to 70 feet (21 meters) tall.

Evolving vascular plants included gymnosperms, a new species of woody plants that relied upon seeds to reproduce. Gymnosperm seeds were better protected for reproductive success than spores, had built-in nutrients to support the plant embryo, and could endure a greater range of temperatures. While some of the earliest vascular plants became extinct in the process of these changes, a few species of gymnosperm called conifers evolved into pine trees that reached 200 feet (61 meters) tall.

Around the time that dinosaurs vanished from the planet, about 120 million years ago, a third major evolution in plants was under way. New plant species called angiosperms appeared. They relied on flowering as part of a new and more effective means of reproduction.

While the gymnosperms simply released their seeds to the air—for instance, in a pinecone—angiosperms produced a flower that was fertilized by an insect or bird. In the process of feeding on the flower's nectar, the visitor carried pollen from one flower to another. Inside the pollen grains were male sperm, and when these sperm were delivered to the female part of the flower, this resulted in the production of fertile seeds that were protected inside a fruit or as part of a nut produced by the plant or tree.

When this fruit was ingested by an animal or a bird, the seeds were carried away. Once released back into the forest as waste, the seed grew to become a baby tree. Nuts also could be carried away and buried by animals for safekeeping; those not dug back up often would grow to become trees. Thus, the angiosperms became the dominant tree species, and they remain so in today's modern forest.

Gymnosperms include the gingko tree (*Ginkgo biloba*) and up to 115 species of pine tree (Genus *Pinus*). Angiosperms make up the largest diversity of land plants estimated at over 250,000 species, including up to 125 species of the maple tree (family *Aceraceae*).

Forest Types

There are different methods to classify a forest—by species, age, ownership, climate, geography, economic or cultural use, or any combination of the above. In the most basic approach, forests can be broken down into two major tree categories: coniferous and deciduous.

The name *conifer* is a Latin word meaning "cone-bearing," and it refers to the conical shell that holds the seeds of these evergreen trees. The name *deciduous* comes from the Latin word *decidere,* meaning "to fall off," and refers to the seasonal loss of leaves that these trees experience at the beginning of winter. While one of these two types of trees may make up or dominate a forest, mixed coniferous and deciduous forests are common worldwide.

Coniferous Trees

Conifers are the oldest and largest trees on Earth. These woody plants grow distinctive needle-like foliage that remains year-round. This type of "needle leaf" species makes up approximately one-third of the total forests on Earth.

The conifers are made up of fourteen different scientific divisions, containing more than 700 species, including cedar, cypress, fir, juniper, pine, redwood, and spruce. Conifer forests occupy up to 31 percent of the nontropical environment, or 4.7 million square miles (12.2 million square kilometers), and can be found across much of the Northern Hemisphere and in parts of the Southern Hemisphere. In general, conifers thrive in colder climates where a short summer season, a thin layer of relatively low-nutrient soil, and moist conditions dominate. Conifers such as spruce and fir are the only tree species that can live in the polar tundra.

Seeds from conifers develop inside a protective pinecone 0.5 to 10 inches (1.3 to 25 centimeters) long and take up to three years to reach maturity. Most seeds are dispersed when the scales on the tree's cone open. Some species require substantial heat from forest fires to force the cone open, keeping seeds protected for up to eighty years. Many conifer species also produce a resin, or sap, as a unique protective medium against insects and birds that prevents their seeds from being entirely consumed.

Deciduous Trees

The remaining two-thirds of the trees in forests have leaves instead of needles. Deciduous forests, also called broadleaf forests, occupy portions of Asia, New Zealand, North and South America, Russia, and other parts of Europe. They make up 65 percent of the world's forests, or 8.9 million square miles (23.1 square kilometers).

Deciduous trees are grouped into four categories according to the leaf arrangements on their branches: single leaves opposite each other (such as oak and maple); single leaves growing alternately (willow and birch); multiple leaves opposite each other (ash and buckeye); and multiple leaves growing alternately (walnut and locust). Up to 2,500 different species of trees exist in deciduous forests.

Most deciduous trees begin their annual cycle by growing new leaves and flowering in the spring. Insects such as bees pollinate trees with flowers and, over the summer months, many of these trees produce a fruit with seeds inside. Others grow seeds in the form of a nut, protected by a pod or another soft or hard casing.

Before or during the pollination process, small green shoots emerge from tree branches and rapidly grow into leaves. Leaves on deciduous trees take many shapes and forms. Some are rounded and smooth, others have fine-toothed indentations along the edges, and still others have a lobe shape. Using photosynthesis, the leaves absorb sunlight and carbon dioxide and release oxygen into the atmosphere.

A few months after a tree's flowers have been pollinated, the tree's produce (seeds, nuts, or fruits) reaches peak size. Much of this bounty is consumed by wildlife and, depending on the tree, by humans. Seeds are spread across the forest floor as they drop to the ground or through animal waste. When conditions are right, these seeds will sprout and, in turn, become new trees.

Examples of leaves and seeds of the two main types of trees are shown here. On the left is a coniferous tree branch with needles and pinecones. At right is a deciduous tree branch with broad leaves and acorns. (© Rob Osman/Fotolia; © Nicolas Gosse/Fotolia)

Winter's inevitable arrival results in deciduous trees going into a state of dormancy. In this process, trees' leaves drop to the forest floor. Over time, this thick layer of organic material decomposes to form rich soil—future food for tree root systems and forest undergrowth.

Forest Biomes

A biome is a large terrestrial area that contains a specific climate and distinctive species. There are six common groupings of forest biomes: boreal forests, temperate deciduous forests, temperate evergreen forests, tropical rain forests, tropical seasonal forests, and savannas.

Boreal Forest

The name *boreal* in Latin means "northern" and includes a large band in the Northern Hemisphere, approximately 6.4 million acres (2.6 million hectares) in size, across North America, Europe, and Asia. These communities, also called Taiga in Asia (from Mongolian), are found in the coldest realms of the world, from Alaska to Siberia to Scandinavia. These locations have short growing seasons followed by long winters.

Boreal forests grow in soils that are high in moisture, thin, rocky, and often low in nutrients. The landscape is dominated by conifer species, including the sitka spruce (*Picea sitchensis*), ponderosa pine (*Pinus ponderosa*), and Siberian fir (*Abies sibirica*). Near the polar regions, the forest height generally is around 10 feet (3 meters) due to thinner soil and very cold weather.

Overall, the boreal forest supports fewer plants and animals than forests in warmer regions. Mammals that are found in this type of forest include the arctic hare (*Lepus timidus*), whose large feet allow it to walk along the surface of deep snow. Another, the arctic wolf (*Canis lupus arctos*), relies on its thick, snowy white coat to remain camouflaged in the tundra. A noteworthy bird living in the boreal forest is the snow goose (*Anser caerulescens*), which feeds in expansive wetlands on plants, grains, and seeds.

Temperate Deciduous Forest

Located in a milder climate than boreal forests, the temperate deciduous forest features four distinct seasons (spring, summer, fall, and winter) with warm summers and cold winters. This kind of forest can be found in the northern latitudes of eastern Asia, eastern North America, and western Europe, covering an area of approximately 5.1 million square miles (13.2 million square kilometers).

This biome is more diverse than the boreal forest, with up to thirty species of common trees, such as the red maple (*Acer rubrum*), black walnut (*Juglans nigra*), and chestnut oak (*Quercus montana*). The temperate deciduous forest receives between 30 and 60 inches (76 and 152 centimeters) of rain per year and has trees that often exceed 120 feet (37 meters) in height.

Many animals that live in this region have adapted to winter conditions by hibernating. Amphibians such as the eastern newt (*Notophthalmus viridescens*) dig into moist soil under layers of leaves to avoid the coldest temperatures. When warm rains thaw the ground in which the newts are buried, they spring to life.

Insects such as the cicada (*Magicicada sp.*) are known for their loud singing pulses and unique life cycle. Dormant for a seventeen-year period when they live underground, one species, the periodical cicada (*Magicicada septendecim*), of these 2-inch-long (5-centimeter-long) insects goes to great lengths to avoid predation.

Temperate Evergreen Forest

The temperate evergreen forest thrives in mild and wet coastal areas along the middle northern and middle southern latitudes, which receive substantial rainfall. Spread across Australia, Central Asia, South America, New Zealand, and western North America, the combination of mountainous terrain and exposure to ocean winds produces humidity between 60 and 80 percent, and these regions often receive more than 100 inches (254 centimeters) of rain per year.

FORESTED AREAS (IN SQUARE MILES) BY REGION AND TYPE

Region	Tropical	Temperate	Total
Africa	2,174,000	19,420	2,193,420
Asia	1,094,000	623,000	1,155,000
Central America	261,000	81,000	342,000
Europe	NA*	700,000	700,000
North America	NA	3,261,000	3,261,000
Oceania	305,000	270,000	575,000
Russia	NA	3,188,000	3,188,000
South America	3,053,000	201,000	3,254,000
TOTAL:	6,887,000	8,343,420	14,668,420

*NA=not applicable.

Source: Convention on Biological Diversity, 2001.

These moderate weather rain forests host a mixture of broadleaf and conifer trees. Resident tree species include the cedar (*Cedrus Var.*), Douglas fir (*Pseudotsuga menziesii*), and coast redwood (*Sequoia sempervirens*).

A number of animal species have evolved to thrive in these moist evergreen forests. The American bald eagle (*Haliaeetus leucocephalus*), whose name means "sea eagle" in Latin, has a wingspan up to 8 feet long (2.4 meters long), which allow it to stay aloft for long periods. The eagle relies on binocular-like eyes and powerful talons to scoop up fish from shallow waters. The sharp-clawed fisher (*Martes pennanti*) resembles a weasel and hunts along the forest floor for small mammals, such as rabbits and squirrels.

The upper canopy of the forest is occupied by the great horned owl (*Bubo virginianus*), which makes large nests in old, hollow evergreen trees. With distinctive, deep vocalizations, these large owls (up to 25 inches, or 63.5 centimeters, tall) are a top predator of the skies, even eating smaller owl species on occasion.

Tropical Rain Forest

Tropical rain forests are found near the equator in Africa, Australia, Central and South America, and on the islands of the southern Pacific where temperatures rarely drop below 68 degrees Fahrenheit (20 degrees Celsius). This combined landmass spans 6.9 million square miles (17.9 million square kilometers) and comprises 5 percent of Earth's land surface. It is characterized by high amounts of rainfall, up to 260 inches (660 centimeters) per year.

Warmth, wetness, and intensely sunlit long days combine to support a concentration of life that is greater than that of the rest of the planet combined. Up to 1.4 million species—or 56 percent of all life on Earth—have been cataloged in this forest habitat. With little seasonal change, the process of a plant going to seed, growing, dying, and decomposing into recyclable organic material happens at a much faster rate than in temperate forests. Trees such as the kapok (*Ceiba petandra*), for example, grow up to 10 feet (3 meters) in a single year, eventually reaching 200 feet (61 meters) in height with a canopy that exceeds 300 feet (91 meters) in diameter.

There is a large diversity of trees in a tropical rain forest. An area less than 1 acre (0.4 hectare) may host 238 tree species, including the Brazil nut tree (*Bertholletia excelsa*) and the fig tree (*Ficus insipida*).

The canopy in a tropical forest is where most mammals live, including the fast-moving spider monkey (*Ateles geoffroyi*), which grows to 20 inches (51 centimeters) tall and weighs less than 12 pounds (5.4 kilograms). This relative of humans has no thumbs but uses its long, strong arms to skillfully navigate across branches and to reach fruit, which it consumes 150 feet (46 meters) off the ground. A common bird species of the tropical forest is the parrot, with over 300 species. One species, the scarlet macaw (*Ara macao*), has brilliant plumage of red, green, yellow, and blue feathers. Macaws in the wild may live up to 50 years and are most common in South America's Amazon basin.

Plants called epiphytes often can be found growing in the tropical canopy. Relying on trees and other plants for structural support, these species of moss, vines, and orchids grow toward the sun to gain energy through photosynthesis. While some epiphytes do not harm their hosts, others, such as the strangler fig (*Ficus barbata*), are parasitic. The strangler fig pushes its roots into the soil at the base of a large tree, climbing upward at the same time, and eventually kills the supporting tree.

Tropical Seasonal Forest

This type of forest can be found across parts of Africa, Central and South America, India, and Southeast Asia. The tropical seasonal forest is considerably drier than a typical tropical one. With less than 90 inches (229 centimeters) of rain per year and two seasons—a dry season (December to April) and a wet season (May to November)—the plants and animals in this ecosystem have adapted to varying conditions.

During the dry season, cooler weather and less moisture prompt trees to drop their leaves. Common trees living in this region include the *indio desnuda* (*Bursera simarouba*) and the foxtail pine (*Pinus balfouriana*). The *indio desnuda* tree (its name means "naked Indian" in Spanish) has a green bark with photosynthetic abilities that helps the tree survive when all of its leaves have dropped. The foxtail pine is one of the longest-surviving plant organisms: It is capable of living 4,500 years.

In the tropical seasonal forest, common honeybees (*Apis mellifera*) store honey to consume during the months when no flowers are present. However, bees still must forage for water to ingest and to cool the hive on warm days.

Reptiles such as the 4-foot-long (1.2-meter-long) ctenosaur (*Ctenosaura similis*) roam the lowland grounds in search of insects, fruits, plants, and small animals. Young ctenosaurs will retreat to the higher trees to flee from predators such as the boa constrictor (*boa constrictor*) snake.

Savanna Forests

Although similar to a seasonal tropical forest, a savanna forest is more open, with fewer trees and bushes. Savanna forests also feature a dry and wet season but receive even less rain—between 10 and 30 inches (25 to 76 centimeters)—per year. Often called dry forests, they can be found in central and northern Africa, Australia, South America, and Southeast Asia and occupy 269,000 square miles (696,710 square kilometers).

Soils in savanna forests are nutrient-poor due to the lack of water and limited organic matter. Trees and plants that thrive in these conditions often are xeromorphic—able to conserve moisture over long periods using different mechanisms, such as spiny leaves that limit their surface area. Trees such as the acacia (*Acacia aneura*) can angle their leaves to reduce direct sunlight and moisture loss. The baobab (*Adansonia digitata*) tree can grow up to 110 feet (34 meters) tall and 35 feet (11 meters) in diameter and, during the dry months, can hold upward of 30,000 gallons (113,562 liters) of water in its trunk and root system.

Fire plays a key role in tree and plant density in the savanna forest. Pine species do not open their cones to release their seeds unless subjected to intense heat from flames. When a fire consumes a portion of a forest, it results in a substantial increase in soil nutrients, allowing new trees and plants to thrive.

Large mammals living in the savanna include antelopes (family *Bovidae*), elephants (family *Elephantidae*), giraffes (*Giraffa camelopardalis*), lions (family *Felidae*), rhinoceros (family *Rhinocerotidae*), and zebras (*Equus zebra*). All of these mammals travel seasonally to stay near water sources in the drier times of the year.

Forest Layers

Depending on the density of the trees and vegetation, each forest contains a defined horizontal stratification. Up to five layers can be measured; these occur at different levels in different climates.

The canopy is the highest part of the forest, found at the treetops. The canopy may reach 250 feet (76 meters) above the ground. The greatest density of tree leaves usually is located here, because this is where the most competition exists for direct sunlight. Bird species are concentrated in the canopy, because it provides safe nesting and roosting locations, as well as good access to food sources.

The understory, which is found one-third of the way down from the canopy, is characterized by smaller trees and less sun. Some vines, clinging onto tree branches, may be able to reach this high in trees, and animals climb into this locale for everything from sleeping to socializing. As this layer receives less sun, there is less vegetation and more open space. The understory's open habitat is excellent for circulating air and often is used by animals as travel lanes.

The shrub layer is located up to 25 feet (7.6 meters) off the ground. This area features woody stemmed plants with trunks measuring up to several inches in diameter. Depending on the amount of sunlight that filters down through the higher layers, the shrub plants either may be sparse (where there is little sun) or densely packed (with full sun). Young trees also may climb upward through this layer.

The forest floor is the second-to-last biotic zone. This area often is the busiest section of a forest, due to plants anchoring their roots and animals living on the ground. Smaller plants, such as grasses, ferns, tree seedlings, shrubs, and bushes, populate this layer and receive intermittent sun during the day. Competition on the ground for moisture and growing space is strong. Many animals and insects live along the forest floor, rather than in the other layers of the canopy, for much of their lives.

Beneath the forest floor is the last forest layer—underground. Often in total darkness, this unseen part of the forest is hidden but not inactive. The immediate subsurface layer contains the newest contributions made by decomposing plant, animal, and insect materials. A variety of species, such as ants, beetles, termites, wasps, and worms, consumes this material and releases a rich waste that resembles finely ground coffee. Just an inch or two lower down is a layer of decomposed soil, which is rich in nitrogen, potash, and phosphorus. This layer is where big and small root systems can absorb water and dissolved nutrients.

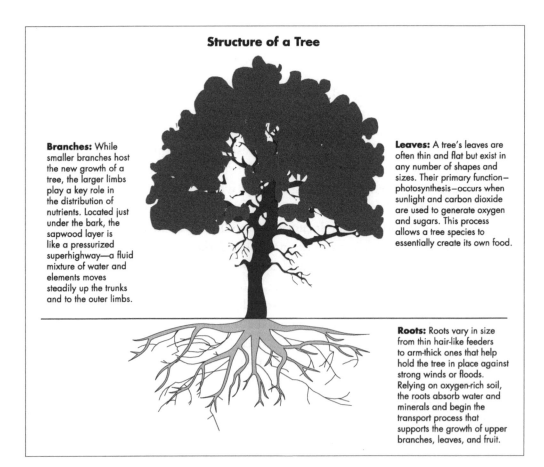

Structure of a Tree

Branches: While smaller branches host the new growth of a tree, the larger limbs play a key role in the distribution of nutrients. Located just under the bark, the sapwood layer is like a pressurized superhighway—a fluid mixture of water and elements moves steadily up the trunks and to the outer limbs.

Leaves: A tree's leaves are often thin and flat but exist in any number of shapes and sizes. Their primary function—photosynthesis—occurs when sunlight and carbon dioxide are used to generate oxygen and sugars. This process allows a tree species to essentially create its own food.

Roots: Roots vary in size from thin hair-like feeders to arm-thick ones that help hold the tree in place against strong winds or floods. Relying on oxygen-rich soil, the roots absorb water and minerals and begin the transport process that supports the growth of upper branches, leaves, and fruit.

Below these thriving layers is the forest base. This thinner but supportive layer is made up of basic organic material, such as woody debris, rocks, clay, and sand.

The Importance of Forests

Forests are complex ecosystems that support a diversity of plants and animals in multiple climate zones across the planet. Forests offer several beneficial ecological services, such as water filtration and oxygen generation.

One particular service is significant land and water protection. The interwoven root systems in a forest not only anchor large trees, they also stabilize the ground, including land covered by smaller trees, bushes, and grasses. This makes the broader area less susceptible to erosion from rain and runoff.

Precipitation in the forest follows three common routes: It is absorbed by forest vegetation; it runs into ponds, lakes, and streams; and it is slowly absorbed into underground aquifers. Forest soils filter water as it travels either to another location or down into an underground water supply.

One of the most noteworthy ecological services a forest provides is its support of biodiversity. Forest ecosystems offer a rich habitat for an array of species, including underground-dwelling beetles, ants building ground nests, fungi growing on tree limbs, and the birds, mammals, and insects living in the canopy, in the tree understory, and on the forest floor.

Biodiversity relies on the nutrient cycle and atmospheric processes. The plant and animal species that occupy this biome participate in several physical and chemical processes that produce nutrients for themselves and other species. These processes capture, convert, and store many elements, including phosphorus and nitrogen, both in the soil and in the plant or animal structure. For instance, when a living organism dies, it is chemically broken down into components that are, in turn, used by other living creatures.

A forest's unique microclimates offer other ecological services. Large trees serve as a windbreak against storms and provide shelter for species living within the forest. The considerable canopy of leaves and branches creates a cooling effect by blocking out and absorbing the sun. The presence of so much plant material reliant on water allows a forest to absorb, in the short term, flooding from storms. The forest canopy also holds light precipitation such as fog and mist. This moisture, which would otherwise be dissipated into the atmosphere, allows organisms to use it to their advantage.

Finally, forests store carbon in living and dead vegetation, as well as in soil. This action, also called a carbon sink, helps to stabilize the Earth's atmosphere by preventing the excessive airborne release of carbon dioxide, a known contributor to climate change. Planting more trees and increasing the coverage of forests has become a common response to abate the impacts of global warming.

Selected Web Sites

Blue Planet Biomes: htpp://www.blueplanetbiomes.org.

Boreal Forest: http://www.borealforest.org.

Radford University Biogeography: http://www.runet.edu/~swoodwar/CLASSES/GEOG235/biogeog.html.

United Nations Food and Agriculture Organization: http://www.fao.org/forestry/en/.

University of California Museum of Paleontology, Forest Biomes: http://www.ucmp.berkeley.edu/exhibits/biomes/forests.php.

Further Reading

Bettinger, Pete. *Forest Management and Planning.* Burlington, MA: Academic Press, 2009.

Davis, Lawrence. *Forest Management.* Long Grove, IL: Waveland, 2005.

Hanson, Arthur, ed. *Our Forests, Our Future: Report of the World Commission on Forests and Sustainable Development.* Cambridge, UK: Cambridge University Press, 1999.

Kimmins, James. *Forest Ecology.* San Francisco: Benjamin Cummins, 2003.

Pearce, David. *The Value of Forest Ecosystems.* Gland, Switzerland: Convention on Biological Diversity, 2001.

Perry, David. *Forest Ecosystems.* Baltimore: Johns Hopkins University Press, 2008.

Young, Raymond. *Introduction to Forest Ecosystem Science and Management.* Hoboken, NJ: John Wiley and Sons, 2003.

2 Humans and Forests

The forest has played a central role in the development of the human species. Dating back 4 million years, this ecosystem has been a vital source of shelter, fuel, food, and freshwater for the first humans who evolved from ape-like ancestors in Africa. At least 30,000 generations before *Homo sapiens* evolved, small bands of people lived in dense tropical forests along locations such as the Tugen Hills of Kenya in east Africa. A pivotal accomplishment included the ability to start their own fires using rocks and sticks.

Migrating northward into Asia and Europe around 1 million years ago, these early peoples found new foods in the temperate forest, such as leafy plants, roots, nuts, fruits, birds, and animals. Small groups of people built basic structures from trees, bark, and leaves. They learned what was edible and planted seeds, developing early agricultural practices.

By 100,000 years ago, humans had evolved to acquire the knowledge of medicine made from various plants and the use of efficient hand tools that were made with rocks and durable wood from the forest. For high-protein nutrition, organized hunting parties used traps and weapons, such as large wooden spears, to capture animals that lived in the forest.

When the most recent ice age ended 12,000 years ago, forested areas regrew and human populations moved into these areas. Bands of Paleo-Indians known as the Clovis peoples settled in the western United States. These groups had a close relationship to the temperate forests in the Rocky Mountain region. Some areas were selectively cleared either by hand or by fire to form cropland for growing vegetables and fruits.

As wood and stone tools improved, technological innovations provided more reliable results, such as the wooden spear with a rock tip called the Clovis point. This well-built spear often was deadly when it struck game animals. Forests also provided materials for the first wooden buildings and boats, which gave the early humans durable shelter and the ability to travel distances across open water.

At the same time, human oral history evolved into drawings, written symbols, and, eventually, written words. The forest featured prominently in traditional stories as the place where humans originated and god-like figures and monsters roamed.

In the last 5,000 years, *Homo sapiens* have harnessed the means to accomplish even greater goals with the forest. Harvesting old-growth trees for wood, finding trees and plants that provide food and medicine, locating and processing minerals and metals, mining underground coal, drilling for and pumping up gas and oil, and utilizing natural water sources collectively anchored the development of human commerce and industry.

Population Growth Impacts Forests

It took humans 3.5 million years to reach a world population of 1 billion people in 1860. Rapid growth since then led the human population to surpass 6 billion in 1997. In 2009, the global population was estimated at 6.7 billion, and it has continued to climb by approximately 82.8 million people each year. By 2050, the world's total population could exceed 9 billion people.

This population explosion has resulted in considerable consumption of natural resources. World economies hunger for the products that forests provide, including land for farming and development, residential and commercial building materials, fuel, and water supplies. Trees are used for cooking and heating and also provide the raw materials for furniture and paper products used in homes, schools, and businesses. Wood by-products include glues, gums, latex, oils, rubber, packaging, plastics, tools, turpentine, waxes, and other everyday items.

The availability of forests and their resources, however, is limited. This is particularly well illustrated in island countries such as Haiti and Madagascar, which have intensely logged their forests over the last 100 years.

Haiti shares a tropical island in the Caribbean Sea with the Dominican Republic. Although geographically connected, the two nations host two completely different environments. While the Dominican Republic has widespread tropical forests, 91 percent of the formerly forested land in Haiti has been denuded.

Haiti has 8.8 million people living in a 10,714-square-mile (26,351-square-kilometer) region; in 2005, the population density was 758 people per square mile. This degree of overpopulation, a high poverty rate, and unregulated deforestation have resulted in extensive damage to the native forests.

A $21 million partnership between the Haitian government and the World Bank began in 1995 to better manage the forest. The project was intended to strengthen the regulatory framework to oversee forests to prevent illegal cutting, but it was halted in 2002 due to lack of local support and limited results. A separate campaign, led by the United States, has resulted in 60 million trees being planted since 1990; however, at least 10 million of these trees have been cut down to make charcoal for sale to residents for cooking. In 2010, a powerful 7.0 magnitude earthquake killed 230,000 and left millions homeless. Haiti already had a limited wood supply for fuel and building reconstruction. In fact, charcoal shortages often led residents to burn trash for cooking fires.

Madagascar has a landmass of 226,500 square miles (586,635 square kilometers) and sits a few hundred miles off the east coast of Africa. In a similar situation to that of Haiti, up to 90 percent of Madagascar's tropical forest has been leveled since the mid-twentieth century. The World Bank participated in an effort with the government of Madagascar to better protect the remaining tropical as well as temperate forests from the economic demands of its population of 19.4 million residents.

Begun in 1975, three projects invested a total of $56 million over twenty-one years. Efforts included establishing tree plantations, replanting cutover areas, implementing new laws and regulations, and educating the public about the importance of maintaining forested areas. Despite numerous setbacks, such as wildfires, and limited citizen participation, the project was completed in 1996 and achieved some success. The island government acted in 2004 to set aside two-thirds of the remaining forest as preserves.

Deforestation

The process of cutting down and removing trees from a forest has been conducted since the first axes were fashioned to chop wood roughly 30,000 years ago. By 2500 B.C.E., farmers in India had developed methods to quickly cut large trees by hand and used oxen to drag the lumber out of the forests to be processed.

In Europe, large-scale logging occurred during two distinct periods. Between 50 C.E. and 500 C.E., the Roman Empire spread across southern and northern Europe. In the process, it removed up to half of the temperate forests, an estimated 468,750 square miles (1.2 million square kilometers), to build structures, heat homes, and support the Roman war machine.

The second period of logging was caused by rapid industrial growth between 1450 and 1950. Countries such as France, Germany, Great Britain, and Spain cut much of their forested land to build commercial mills, operate large

Deforestation, whether achieved through the logging or burning of forests, affects the surrounding area in many ways. It can lead to desertification, changes in local weather patterns, the extinction of plant and animal species, and displacement of indigenous peoples who previously relied on the forest's resources. (© Mark Atkins/Fotolia)

boat fleets, manufacture industrial products, and heat the homes of a growing population. More than half of the central European forests were harvested to support these burgeoning economies.

The settlement of North America and the subsequent creation of the United States started an intense period of logging that would last more than 100 years and cut 200 million acres (81 million hectares) of temperate forest. By 1700, the nation's northeastern coastal regions had been heavily deforested both for wood and to clear the land for new farms.

The eastern cities of Baltimore, Boston, New York, and Washington, D.C., required enormous amounts of lumber for new buildings. Logging crews moved throughout southeastern Canada, Maine, New Hampshire, and Vermont in order to supply wood to the growing cities. By 1810, much of the region had been exhausted of its oldest and biggest trees, and a westward movement began.

By 1820, the Great Lakes region's old-growth stands of spruce and pine trees were supplying the needs of the developing nation. By 1850, larger communities were well established both in the region and farther west, including Chicago, Dallas, Denver, San Francisco, Seattle, and St. Louis.

The Midwest was by then largely cutover, so attention turned to the Pacific Northwest forests. There, Douglas fir trees, up to 11 feet (3.4 meters) thick were able to produce eight 32-foot-long (9.8-meter-long) logs. This region would support ninety years of uninterrupted logging over three generations due to the larger tracts of forests and their considerable size.

In the 1880s, new coal-powered and steam-driven tractors increased the pace of logging. The advent of equipment such as gas-powered chainsaws, as well as bulldozers and other large hydraulic equipment, again leapfrogged the rate at which trees could be cut and removed.

The widespread deforestation of the American landscape that resulted was responded to with new federal forestry laws and the creation of a national forest system. Efforts to balance conservation of forest tracts with selected removals took decades to implement. By the 1980s, most large-scale logging operations in the West finally were replaced by managed forestry and the logging of smaller areas.

While the pace of temperate deforestation slowed in the twentieth century, worldwide logging efforts during this century were the largest ever. This was driven by the desire for low-cost forestry products to supply the global market. This demand has been met by international corporations that buy a tract of land in a remote area where it is inexpensive and few regulations exist, log the land, and then resell the property to another industry, for farming or mining. This allows the corporation to reap the immediate value and then redirect its investment to other tracts for the next tree harvest.

Advances in machinery have allowed for tropical forests in Africa, Asia, the Pacific Islands, and South America to be quickly logged. Today, a crew of ten

ANNUAL PAPER CONSUMPTION

Country	Millions of Metric Tons
Canada	7.4
China	36.2
France	11.3
Germany	19.1
Italy	10.9
Japan	31.7
United Kingdom	12.6
United States	92.3

Source: United Nations Food and Agriculture Organization, 2002.

with the newest machines can harvest the trees from 35 acres (14.2 hectares) of forest in a single day.

The practice of buying up large tracts of land, logging them, and then selling them first became common in the 1980s. This resulted in a removal rate of 37.7 million acres (15.3 million hectares) worldwide per year, up to 1 percent of the total tropical forest area, between 1990 and 1995. In South America's Amazon basin, up to 9,062 square miles (23,471 square kilometers) were cut in 2003 alone.

In another part of the world, the Philippines and Thailand have cut half of their mangrove forests since 1960. These wetland forests have been converted into farms, especially for fish and shrimp export. Much of the mangrove wood is cut into firewood and sold for residential cooking needs.

State of World Forests

There are an estimated 13.8 million square miles (35.7 million square kilometers) of forest on the planet, which are found on every continent except Antarctica. Sixty-six percent of worldwide forests are found in seven countries: Brazil, Canada, China, Democratic Republic of the Congo, Indonesia, Russia, and the United States. Twenty-nine countries are 50 percent forested, with twenty-one of these countries located in the tropics. Forty-nine countries have less than one-tenth of their landmass covered by forests.

These diverse forestlands have provided a bounty of resources for human consumption. The forest community provides wood for fuel to heat homes, cook food, and produce electricity. The biome is used as a source of both animal and plant foods. Some isolated communities rely entirely on a local forest for all of their food, energy, tools, and medicinal needs.

Despite these benefits, forest overuse is a chronic problem. Human population growth and international trade have resulted in a substantial worldwide depletion of forested areas. In 1800, up to 44 percent of the dry land was forested. This number shrank to 39 percent by 1990 and, by 2008, was estimated at 28 percent. Globally, up to 27.1 million acres (11 million hectares) of forest have been clear-cut each year since 1990.

Recognition of the unsustainable pace of forest removals began with a series of United Nations studies during the 1990s. Despite a significant attempt to educate the public and better manage public forests around the world, up to 91 million acres (37 million hectares) were heavily cutover between 2001 and 2006. Although reforestation efforts have resulted in new forests being planted at a rate of 3.9 million acres (1.6 million hectares) per year since 2001, only about 10 percent of what is removed annually is replaced.

CHANGES IN WORLD FOREST COVER

Continent	1990 Forests (Million Acres)	2005 Forests (Million Acres)	Percentage Change
Africa	172	156	−9.30
Asia	141	140	−0.70
Europe	244	246	0.82
North and Central America	175	174	−0.57
Oceania	52	51	−1.92
South America	220	210	−4.54
World Totals:	1,004	977	2.69

Source: Global Forest Resources Assessment, United Nations Food and Agriculture Organization, 2005.

Continents with alarming rates of deforestation include Africa, which lost 4.3 percent of its forest, or 16 million acres (6.5 million hectares), between 1990 and 2005, as industrial logging surged and wood was cut for heating and cooking by local residents. South American forests also were heavily cutover, with 10 million acres (4 million hectares) removed during the same fifteen-year period. Only Europe as a region gained forested areas, with a 2 million acre (800,000 hectare) increase achieved through a combination of European countries relying on other regions of the planet for forest resources and extensive environmental restoration programs.

Forest Peoples

Up to 1.4 million people live in forest communities mostly in Africa, Asia, the Pacific Islands, and South America. They rely on the tropical forest for their livelihood, and isolated populations may have little contact with outsiders. These "forest peoples" generally use the forest as a source of food and materials to make their own shelter, clothing, medicines, and other necessities. In some cases, they are fiercely protective of their territory and unique cultures.

In Africa's Congo region, for example, the Bambuti people live in small groups of ten to sixty-five that hunt and gather in the Ituri Forest (24,324 square miles, or 62,999 kilometers, about the size of West Virginia). The Bambuti are pygmies and therefore have small bodies, averaging only 4 feet 10 inches (1.5 meters) tall. They roam great distances for food, relying on a diverse diet that includes fruits, nuts, berries, roots, honey, wild pig, fish, and shellfish.

This formerly deforested area, in the Sierra de Aracena mountain region of Andalusia in southern Spain, has been replanted with new young trees. *(Chris Sattlberger/Image Bank/Getty Images)*

Another forest people, the Guaica Indians, live in Brazil in the Amazon Rain Forest at the headwaters of the Orinoco River. They number around 20,000 and are spread out over hundreds of square miles. The traditional Guaica live together in small family groups that share food and resources. Men, who traditionally shave their heads and use purple dye on their upper bodies for camouflage, hunt in groups with bows and arrows, while women maintain community areas, tending to the banana, yam, cotton, and tobacco crops.

Forests to Farms

As larger human populations grew in Asia, Europe, and North America in the eighteenth century and then in Africa, Central and South America, and the Pacific Islands in the twentieth century, the need for pastures and croplands expanded. More than 200 years ago, worldwide uplands were divided into three groups: forests (35 percent), fields (35 percent), and rock/tundra/desert (30 percent). In 2008, forests had shrunk to 28 percent, while fields and rock/tundra/desert each grew to 40 and 32 percent, respectively.

Although natural grasslands were easy to convert to farms, most locations away from floodplains lacked the rich soil necessary for growing crops. This resulted in large tracts of forest being cleared; the wood was used to construct buildings and for fuel. Between 1850 and 1990, up to 3.2 million square miles (8.3 million square kilometers) of temperate and boreal forests were cut down to create animal pastures and cropland.

Since 1970, the most deforestation has occurred in the tropics, where 2.3 million square miles (6 million square kilometers) of forest have been cleared. An estimated 27 percent of forested land was altered for crops and 18 percent for livestock, leaving 55 percent of the deforested tropical land for residential, commercial, or industrial uses. Half of the worldwide croplands was added in the last fifty years since the mid-twentieth century, demonstrating the rapid rate of conversion.

Modern industrial farming, as it is most commonly practiced, can have an exhaustive effect on the soil. Nutrients such as phosphorus, nitrogen, and potash may be high when a forest is first cut, but they can be quickly depleted by nutrient-hungry crops such as soybeans. Methods such as crop rotation and organic practices have proven to lessen the impact.

Once an area is cutover, other problems can be created without careful management (discussed below). These problems include soil erosion from rain runoff, soil and water pollution from the use of pesticides and herbicides, as well as reduced capacity of the soil to absorb and temporarily hold water. Desert-like conditions also are common in deforested areas, as large-scale tree removal results in an increase in the local temperature and fewer rainstorms.

If crops are not properly managed through sustainable agricultural practices, farm productivity rapidly decreases two years after trees are removed; if soils are not enriched, the land becomes unproductive. Up to half of the land historically cleared for farming has been abandoned. Although some of it becomes developed into residential or commercial tracts, the majority is allowed to return to a forested state, albeit in a denuded, nutrient-depleted, topographically altered environment. In such a landscape, invasive species often flourish, and a healthy forest may take a century to become reestablished.

Forests and Carbon

One of the greatest assets of a forest is its ability to store the element carbon. A forest is referred to as a carbon sink, because it soaks up more carbon than it emits. By using photosynthesis (taking in carbon dioxide and releasing oxygen) in the surfaces of vegetation, forest ecosystems store carbon in their bark, branches, leaves, roots, and soil. The element is then broken down into soil layers when vegetation decays.

Forests can hold this carbon for more than 3,000 years, releasing only a small portion of it back into the atmosphere at any one time. This process prevents a buildup of atmospheric greenhouse gases, including methane and nitrous oxide. Forests today hold up to 283 gigatons (1 billion tons = 1 gigaton) of carbon in their vegetation (leaves and stems), 38 gigatons in dead wood, and 317 gigatons in soils. This total of 638 gigatons held by the forests is a significant number, although 37,000 gigatons of carbon is held in the oceans.

Steady deforestation over the last 300 years has resulted in large amounts of carbon being released into the environment, contributing to an increase in the estimated atmospheric concentration of carbon dioxide from 278 parts per million in 1750 to 385 parts per million in 2009. Deforestation has resulted in up to 1.6 billion tons of additional carbon being released from forests and transported into the air each year, which is up to one-fifth of the annual carbon emissions worldwide.

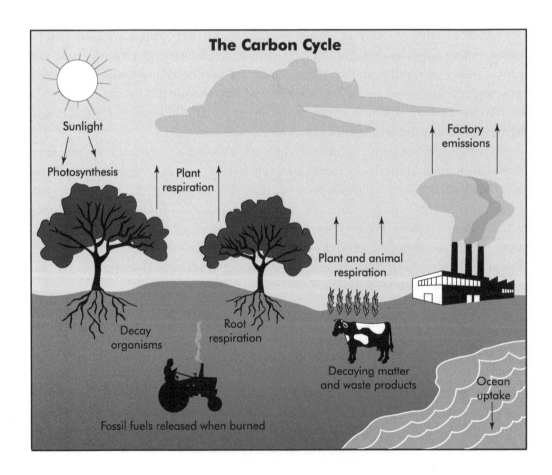

CARBON LOSSES FROM DISTURBED TROPICAL FORESTS

Type of Disturbance	Percentage of Carbon Lost to Atmosphere
Animal Pasture	90 to 100
Planted Crops	90 to 100
Degraded Forest*	20 to 50
Logging	20 to 40
Tree Plantations	30 to 50

*"Degraded" means logged and/or cutover.

Source: George Woodwell. *Forests in a Full World.* New Haven, CT: Yale University Press, 2001.

Tropical regions have faced the highest yearly forest removals. Regions affected include Indonesia (7,335 square miles, or 18,998 square kilometers) with forest removals for pulp and palm oil, Sub-Saharan Africa (15,444 square miles, or 40,000 square kilometers) for charcoal and building materials, and South America (16,602 square miles, or 42,999 square kilometers) for wood and farming.

With total forest removals exceeding 50,193 square miles (130,000 square kilometers) per year, in the 1990s, the international community began working through the United Nations to reduce this loss by better protecting forest areas. One measure being taken is ensuring that cutover regions are replanted right away to stem large carbon emissions.

Managing Forests

Forest conservation dates back thousands of years to early human settlements. Smaller communities that relied on the forest developed practices of conserving resources to preserve forested lands for long-term use. Although these rules seldom were written down, they were passed along from generation to generation.

For example, for hundreds of years the Haida Native Americans in southeast Alaska have handpicked old-growth sitka spruce (*Picea sitchensis*) and red cedar (*Thuja plicata*) trees to be carved into large canoes. As a practice, they would travel great distances to use trees from varied locations. If a tree was not yet ready to be harvested, the Haida would continue on.

As the human population grew in Europe, wood consumption produced wide swaths of deforested regions. The landscape changed from one dominated by forests to one where grasslands and prairie controlled the view for miles.

As early as 1086, the British government, led by William of Normandy, ordered a detailed survey of all commonwealth lands to assess the extent of forested areas. This survey proved that forests had been greatly reduced in size. William directed his officials to require more tree planting efforts and restricted harvests within any public forests until such forests had reached mature size with large trees.

In Japan, up to 11,583 square miles (30,000 square kilometers) of forest were decimated in the 1640s to build several large castles, causing widespread protest among the citizens. In response to demanded reforms, Lord Matsudaira Sadatsuna declared in 1650, "For every tree felled, plant 1,000 seedlings." The Japanese word *kousei* meaning "regeneration" became a central concept in forest management across the island country.

In 1662, an Englishman named John Evelyn appeared before the Royal Society of London to address the lack of hardwoods for shipbuilding and the need to plant trees to rejuvenate the decimated forests. His book *Sylva: Or, a Discourse of Forest Trees* served as a benchmark for the emergence of the field of professional forestry.

By 1750, British government employees who had lived in Asia and learned how to replant cutover tracts of trees with preferred species brought back their knowledge to be used in the English countryside. Alexander von Humboldt, a German scientist and explorer, supported methods of forest conservation at the turn of the nineteenth century. Humboldt observed the results of South American tropical forests being cut selectively, and he urged German politicians to protect their old-growth forests.

The rapid growth and expansion of the United States between 1700 and 1900 corresponded with an unprecedented removal of temperate forests. George Perkins Marsh's *Man and Nature* (1864) was a groundbreaking modern ecological publication that outlined the impacts of humans on the environment. One of the book's arguments—that civilizations such as Greece and Rome fell due to extensive environmental degradation—gained widespread attention.

In response to public concern, President Benjamin Harrison signed the Forest Reserves Act in 1891. That same year, 13 million acres (5.3 million hectares) of land were protected as the first national forests. By 1897, the U.S. Forest Service Organic Administration Act was signed into law. It set the stage for the creation of the U.S. Forest Service in 1905.

This new agency's mission was "to sustain the health, diversity, and productivity of the nation's forests and grasslands to meet the needs of present and future generations." By 1907, the agency oversaw 56 million acres (23 million hectares) of forest, and by 1910, 150 national forests had been created, covering 172 million acres (70 million hectares). One hundred years later, in 2007, the U.S. Forest Service had grown to 30,000 employees overseeing 155 national forests and twenty grasslands, and managing 193 million acres (78 million hectares).

In tropical areas of the world, forest protection efforts have developed more recently. Africa, Asia, the Pacific Islands, and Central and South America host most of the world's tropical forests and have growing economies reliant on the forest products industry. Nations in such areas often are in debt to larger industrial nations, and cutting forests to generate lumber or to convert the land to agriculture is a quick way to increase gross national products. In the last decades, the most heavily logged nations have included Brazil, the Democratic Republic of the Congo, Indonesia, Mexico, Sudan, and Zambia.

Since the 1970s, a number of international programs—offered by groups such as The World Bank, the United Nations, and the World Wildlife Fund— have attempted to change forest management practices in these heavily logged regions. Emphasis has been placed on sustainable economic measures, such as encouraging ecotourism and agroforestry (an integrated approach to combining the management of forests with growing crops or raising livestock), replanting cutover areas, establishing forestry education programs, and improving forest management efforts.

Zambia's forests in Southern Africa make up 60 percent of the upland habitat, spanning some 247,105 square miles (640,000 square kilometers). Unregulated tree removals of up to 3,281 square miles (8,498 square hectares) a year between

THE LARGEST U.S. NATIONAL FOREST

The largest publicly owned forest in the United States is located in southeast Alaska in the Tongass National Forest. The Tongass consists of 26,250 square miles (67,988 square kilometers)—about the size of West Virginia—including 11,000 miles (17,699 kilometers) of shoreline.

This temperate rain forest is a veritable island, as it is surrounded by ocean waters, glaciers, and rocky peaks. However, because the Tongass has been a major logging site since the 1950s, only 30 percent of uncut old-growth forest remains, with some mature stands estimated to be 500 years old. The forest is home to a wide variety of wildlife, including the largest brown bear and bald eagle populations in North America.

1990 and 2001 led to calls for conservation. In the last decade, the Zambian government has created a total of 481 protected areas consisting of 300 local forest reserves and 181 national forests. However, despite a new regulatory agency and better enforcement of government-protected forests, illegal cutting persists. This is due to widespread poverty, lack of local government support of conservation laws, and thriving wild game and firewood markets.

Selected Web Sites

European Forest Institute: http://www.efi.int/portal/.
Forest Peoples Programme: http//www.forestpeoples.org.
Global Forest Watch: http://www.globalforestwatch.org/english/index.htm.
U.S. Forest Service: http://www.fs.fed.us.
United Nations Food and Agriculture Organization, State of the World's Forests: http://www.fao.org/forestry/site/sofo/en/.
World Wildlife Fund (United States site): http://www.worldwildlife.org.
WWF–World Wide Fund for Nature (global site): http://www.panda.org.

Further Reading

Allaby, Michael. *Temperate Forests*. New York: Facts on File, 1999.
———. *Tropical Forests*. New York: Chelsea House, 2006.
Evelyn, John. *Sylva: Or, a Discourse of Forest Trees*. 1662. Whitefish, MT: Kessinger, 2007.
Gibson, Clark, ed. *People and Forests*. Cambridge, MA: MIT Press, 2000.
Honnay, Olivier, ed. *Forest Biodiversity*. Oxfordshire, UK: CABI, 2004.
Hunter, Malcom. *Maintaining Biodiversity in Forest Ecosystems*. Cambridge, UK: Cambridge University Press, 1999.
Marsh, George Perkins. *Man and Nature*. 1864. Seattle: University of Washington Press, 2003.
Montagnini, Florencia. *Tropical Forest Ecology*. New York: Springer, 2005.
Starr, Chris. *Woodland Management*. Wiltshire, UK: Crowood, 2005.
Woodwell, George. *Forests in a Full World*. New Haven, CT: Yale University Press, 2001.

FORESTS OF THE WORLD
CASE STUDIES

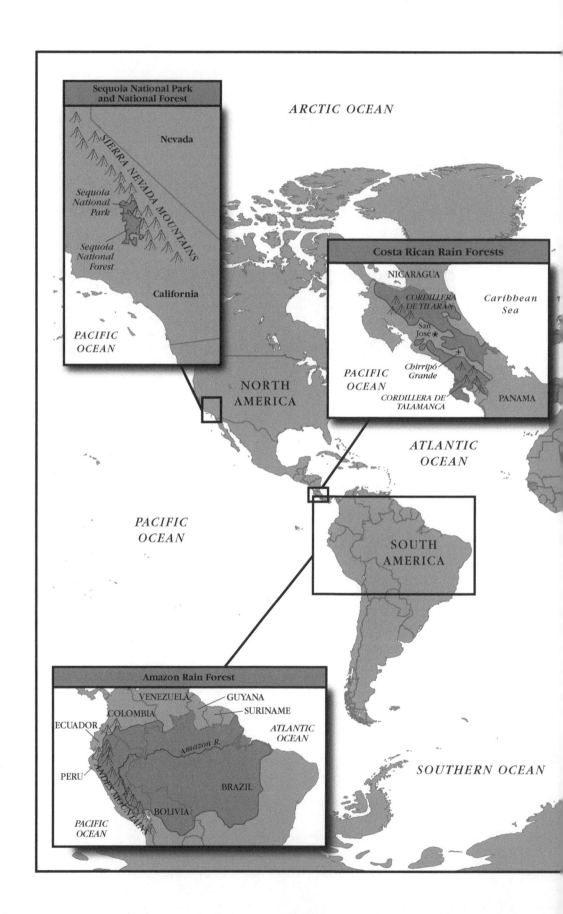

Sequoia National Park and National Forest

Nevada

SIERRA NEVADA MOUNTAINS

Sequoia
National
Park

Sequoia
National
Forest

California

*PACIFIC
OCEAN*

ARCTIC OCEAN

Costa Rican Rain Forests

NICARAGUA

*CORDILLERA
DE TILARÁN*

*Caribbean
Sea*

San
José✳

*PACIFIC
OCEAN*

*Chirripó
Grande*

*CORDILLERA DE
TALAMANCA*

PANAMA

NORTH
AMERICA

*ATLANTIC
OCEAN*

*PACIFIC
OCEAN*

SOUTH
AMERICA

Amazon Rain Forest

VENEZUELA — GUYANA

COLOMBIA — SURINAME

ECUADOR

*ATLANTIC
OCEAN*

Amazon R.

ANDES MOUNTAINS

PERU

BRAZIL

BOLIVIA

*PACIFIC
OCEAN*

SOUTHERN OCEAN

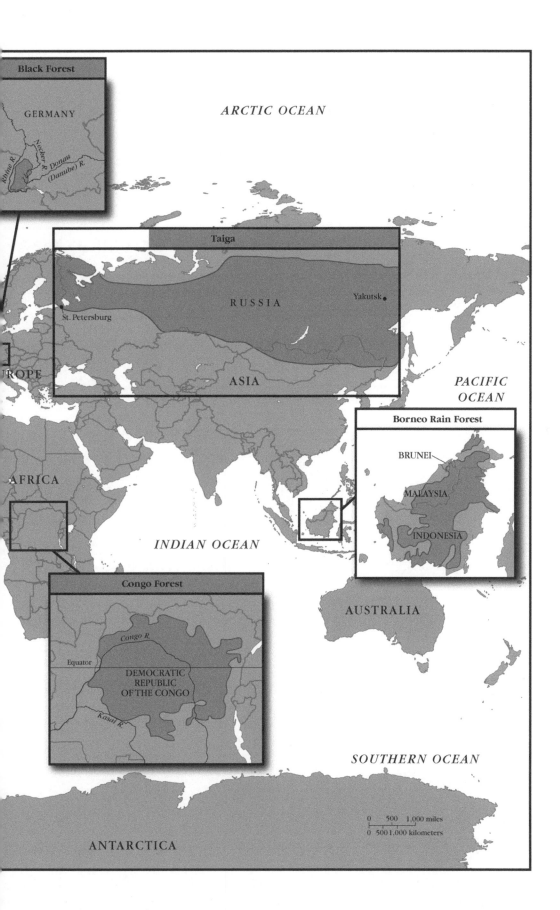

3 Amazon Rain Forest Brazil

The world's largest rain forest spans 2.3 million square miles (6 million square kilometers) across the equatorial region of South America. This expanse, known as the Amazon Rain Forest or Amazonia, is about the size of the continent of Australia.

The forest is spread across 39 percent of South America and includes parts of eight countries: Bolivia, Brazil (which holds 61 percent), Colombia, Ecuador, Guyana, Peru, Suriname, and Venezuela. With a warm and moist environment that ranges from 68 to 84 degrees Fahrenheit (20 to 29 degrees Celsius), most regions of the forest receive up to 8.5 feet, or 102 inches (259 centimeters), of rain annually.

The forest contains the second-longest river on Earth, the 4,180-mile-long (6,726-kilometer-long) Amazon River, and its tributaries. This ecosystem is so productive that its trees and plants generate up to one-fifth of Earth's oxygen—and thus is known as the "lungs of the world."

The rain forest also has one of the highest densities of living organisms. An estimated 1.5 million species reside here, including more than 3,500 species of plants, 2,500 species of trees, 600 species of birds, 425 species of mammals, 440 species of amphibians, 375 species of reptiles, and 260 species of fish.

The Amazon River begins in the Andes Mountains in central Peru, 130 miles (209 kilometers) east of the Pacific Ocean. Originating in snowfields at 13,000 feet (3,962 meters) above sea level, this body of freshwater has an average annual flow rate at its mouth where it empties into the Atlantic Ocean of 7.7 million cubic feet (218,040 cubic meters) per second. It relies on a maze of several thousand smaller rivers and tributaries that link multiple watersheds.

The Amazon basin holds up to 19 percent of all the surface freshwater on the planet. Where the river meets the Atlantic Ocean, it spans 247 miles (397 kilometers) in width. Freshwater from the Amazon dilutes the salty ocean waters and influences coastal currents more than 90 miles (145 kilometers) offshore.

The Amazon forest is relatively new in terms of geologic history. Roughly 145 million years ago, the continents of South America and Africa separated during the Cretaceous Period due to tectonic plate movements, resulting in the formation of the Atlantic Ocean. A mountain ridge called the Purus Arch formed on the eastern side of present-day Brazil. This redirected Amazon River water flows from east to west, toward the Pacific Ocean.

When the Andes Mountains formed around 65 million years ago, the direction of water in the Amazon River switched directions to flow east again. However, the Purus Arch still existed and served as a dam for a portion of the eastward flowing water. These events resulted in considerable flooding of the Amazon forest valleys, forcing species onto higher terrain "island communities" for thousands of years. Once the Purus Arch eroded away, water levels shrunk to a single river flow and the forest re-emerged in the lowlands.

The Amazon forest is a fragile ecosystem. Because only the top few inches of soil contain suitable nutrients for plant growth, most trees access nutrient-rich soil through a complex network of thick, buttressed roots that can extend 70 feet (21 meters) outward. The larger trees grow up to 130 feet (40 meters) high, forming a tight canopy of leaves that allows little direct sunlight to pass through to the ground. Physical adaptations for survival by trees in this unique ecosystem are demonstrated by species such as the capirona (*Calycophyllum sp.*), a rot-resistant hardwood that can live 750 years.

Another rain forest giant is the kapok tree (*Ceiba petandra*), which is capable of reaching 150 feet (46 meters) skyward and often is covered with several thousand pounds of vines and climbing plants, called epiphytes. Scientists have cataloged 2,500 species of trees in the Amazon, which may be less than half the number of species that exist in this forest.

Rain forest tree species do not have easy-to-read growth rings. This is due to the fact that the growth seasons are not separated by periods of inactivity, which account for the clear growth rings formed in trees growing in temperate forests.

Animals in the Amazon thrive in various and diverse habitats. The giant river otter (*Pteronura brasiliensis*) is almost 6 feet (1.8 meters) long and weighs up to 65 pounds (29.5 kilograms). It lives along riverbanks and flooded fields; its diet consists of mostly perch and catfish. Highly vocal, it makes nine distinctive sounds. These otters are easy to approach, and poaching for their pelts has reduced the population to 5,000, making them endangered.

Bird species live at every level of the Amazon Rain Forest. The Spix's Guan (*Penelope jacquacu*) prefers the lowland rain forest areas with less dense stands of trees. Nearly the size of a domestic chicken, it measures about 3 feet (1 meter) from head to tail and lives most of its life above ground in the middle of the forest canopy, eating plants, seeds, and fruits, depending on the season. This affords the species considerable protection from predators roaming the forest floor.

Another distinctive animal is the Brazilian pink tarantula (*Vitalius platyomma*), which thrives in the warm and moist jungle. Almost 9 inches (23 centimeters) in diameter, this spider is curious but nervous and can hurl small, sharp hairs at a predator, causing painful irritation. A common myth about the nearly 900 species of tarantula that live worldwide is that their venomous bite is fatal for humans. An adult tarantula can kill small lizards or mice but not humans.

The tropical forest floor hosts a constant cycling of nutrients among the animal and plant communities. Leaves drop year-round in a slow but steady rate. Insects or animals consume most leaves, and others decompose within weeks.

Key to the environment are Amazonia's forty-one species of leaf-cutter ants (*Acromyrmex sp.*), which live in colonies of up to 3 million ants. Individuals use their strong scissor-like appendage to slice segments of leaves and carry

Leaf-cutter ants (*Acromyrmex sp.*), which form underground colonies of millions of individuals, use cut leaves to grow a *Lepiotaceae* fungus, which they consume. These social insect species are common in the rain forest habitat. (© Ismael M Verdu/Fotolia)

them underground. Once there, the pieces support the growth of *Lepiotaceae* fungus, which the ants consume. Working in large groups, leaf-cutters blend into their landscape. Even a thousand bits of bright green leaf moving along a trail may be camouflaged on the forest floor.

Due to the rapid rate of consumption and decay of plant matter, the Amazon forest soil, or humus, does not accumulate the way forest soil in temperate climates does. Rain forest soils are relatively thin—only a few inches—and are leached of most nutrients by plant consumption and steady rains. They often are reddish in color, which means they are rich in iron oxides and other metals, derived from the underlying metamorphic rock beds.

Human Uses

Around 12,000 B.C.E., Asian peoples crossed the Bering Strait land bridge between Russia and North America from Mongolia into Alaska and traveled south. Many generations later, they settled in the Amazon forest around 6,000 B.C.E. The forest population density is now 1.5 people per square mile.

Since European contact, the indigenous population of the Amazon has been reduced by 50 percent due to introduced disease, armed conflict, and disrupted food supplies. The Brazilian portion of the Amazon contains up to 170 partially nomadic indigenous tribes, numbering around 535,000 in 2009. Many of these native peoples rely on subsistence fishing, hunting, farming, and ranching. As part of the recognition of the rights of the native population, the Brazilian government has designated 26.4 percent of Brazil's Amazon for the indigenous population but has provided little funding for public projects such as roads, schools, health care facilities, and utilities.

One of the indigenous peoples living in the Amazon is the Yanomami. These native Brazilians are spread across a region spanning 780,000 square miles (2.02 million square kilometers). Groups range in size from a few dozen up to 350 people. The Yanomami had little contact with outsiders until the 1970s, when gold miners and loggers built roads into their territory. Soon, the Yanomami were exposed to a number of diseases, including tuberculosis, measles, and the flu. The population plummeted from around 75,000 to an estimated 32,000 by 2008.

Violence sometimes erupts between the Yanomami and outsiders, especially when timber has been cut illegally on the natives' claimed lands. Most live by their tribal ways, subsisting solely on forest products, but others have acquired motorized vehicles, processed foods, and guns.

Nonnative Brazilians have encroached upon the Amazon steadily since the 1960s. Their settlements, concentrated in the southern state of Mato

Grosso and eastern state of Para, exceeded 13.5 million people as of 2008. The Amazon's urban growth rate was 3.5 percent in 2007.

Most Amazon residents are employed in one of three industries: logging, farming, or ranching. Brazil has become the world's largest exporter of beef (2.2 million tons in 2008) and the second-largest exporter of soy (46 million tons in 2008). Farms account for up to 13 percent or 429,000 square miles (1.1 million square kilometers) of Amazonia—about the size of the nation of South Africa.

Pollution and Damage

In the last four decades of the twentieth century and into the twenty-first, the Amazon Rain Forest ecosystem has lost nearly 30 percent of its forest habitat due to population growth. Between 2000 and 2006, 64,637 square miles (167,410 square kilometers)—an area larger than the state of Georgia—was deforested. At this rate, up to 42 percent of the Amazon could be deforested by 2020.

FOREST PEOPLES FROM AROUND THE WORLD

- More than 1.6 billion people depend on the forest for a portion of their income.

- Nearly 1.2 billion people use elements of agroforestry to grow food.

- Up to 350 million people live adjacent to a forest used for daily subsistence such as food and fuel.

- More than 60 million people are completely indigenous, living in and relying solely on forests for their needs.

Source: Annual Report, Forest Peoples Program, 2008, and The World Bank, 2004.

This record of resource consumption dates to 1964, when the military took control of the Brazilian government (it held power from 1964 to 1985). The generals saw a nation of 70 million people living in urban centers, while up to half the country was composed of rich forest that held only 2.4 million residents.

In an unprecedented action, the military government built the 1,980-mile (3,186-kilometer) Trans-Amazon Highway and offered settlers free 250-acre (101-hectare) lots. The project took twenty years to complete and was plagued with problems. The road was susceptible to floods and washouts, despite tons of stone and occasional pavement. The disturbance brought new species to the dense jungle, with coastal mosquitoes spreading malaria across the forest and killing thousands of indigenous people.

By 1981, up to 55,000 people had moved into the valleys of the Amazon Rain Forest, settling on land located near the new highway; however, within five years, 60 percent of the settlers had returned home. The so-called Plan for National Integration was deemed a failure. Since 1985, a democratic government has been in place in Brazil.

The Trans-Amazon Highway set the stage for lawlessness and illegal development that continues into the twenty-first century. Loggers, miners, and business developers throughout the forest region have built up to 109,000 miles (175,381 kilometers) of illegal roads. Once a primitive road is built, a logger will clear-cut the accessible land. Within a year, squatting farmers will arrive, burn the remaining debris to clear the land, and plant crops. Up to 62,000 false titles to land have been sold to unknowing buyers, resulting in voided titles by government authorities.

Enforcement and oversight in the Amazon are insufficient, and soldiers, police, and other governmental regulatory agents are underpaid, making corruption widespread. Land conflicts in the state of Para alone have resulted in armed violence that has claimed the lives of more than 820 residents since 1997.

The deforestation of the Amazon has brought unforeseen environmental problems to Brazil. Tree removal created an arid environment and exposed land directly to the tropical sun; air temperatures in the exposed land are 20 degrees Fahrenheit, or 6.7 degrees Celsius, hotter than in the forest. In addition, the level of oxygen production declines due to fewer plants.

Because the presence of a large forest results in considerable moisture held in vegetation, soils, and water bodies, the Amazon supplies the water for up to 40 percent of area precipitation. In the last decade, the loss of forests has contributed to a number of widespread droughts. One drought in 2004 resulted in Amazon River levels dropping 45 feet (14 meters), leaving previously riverside communities parched and up to 2 miles (3.2 kilometers) from the river's edge.

AMAZON DEFORESTATION

Year	Deforestation (Square Miles)	Year	Deforestation (Square Miles)
1978–1988	8,158	1999	6,663
1989	6,944	2000	7,658
1990	5,332	2001	7,027
1991	4,297	2002	9,845
1992	5,322	2003	9,500
1993	5,950	2004	10,088
1994	5,751	2005	11,745
1995	11,219	2006	8,774
1996	7,013	2007	4,490
1997	5,034	2008	4,984
1998	6,501	2009	3,860*
		TOTAL:	142,821**

*2009 figure is an estimate as of late 2009.
**Total square miles of deforestation nearly equal the size of the state of California.

Source: Brazilian National Institute for Space Research, 2009.

Deforestation also has made the region susceptible to considerable erosion. When this happens, plants and trees cannot reestablish in the scarred areas, leaving the land barren.

Mitigation and Management

The widespread deforestation of the Amazon came into the international spotlight in the 1990s. Evidence of clear-cuts and burning forests broadcast by news outlets and displayed by published satellite imagery raised considerable public alarm. Environmental groups blamed corrupt governments that employed a frontier-like mentality and ignored complex land ownership issues.

Brazil, due to its large population, was looked to for better forest management leadership. A number of land-use reform programs became the focus of the federal government's conservation efforts. In 2000, it initiated a plan to limit forest logging and auctioned off temporary leases for selective logging in exchange for royalty fees. The courts and government also found a large number of land titles invalid in 2003, and the authorities repossessed land as publicly held. In 2007, up to 150 new employees were hired to oversee a federal forest services department.

In 2003, a unique accord called the Amazon Region Protected Areas Program was developed between the Brazilian government and the World Wildlife Fund (WWF). With the intention of conserving national parks and reserves, the WWF provided funds to help manage and enforce existing laws.

By 2012, this accord plans to protect a forest area larger than the state of California, relying on a $390 million endowment from environmental and government groups. Some Brazilian leaders, who claim these actions are part of environmentalists' effort to take over the Amazon for their own use, have bitterly criticized the program.

Efforts to better protect the Amazon forest have been led by the Brazil Ministry of the Environment. During 2006, the ministry conducted 134 major operations to combat deforestation in the Amazon. These included halting the transport and sale of illegal lumber, inspecting logging operations, and educating local communities about forest protection laws and regulations. This project was carried out by a staff of 3,102, resulting in 5,745 fines totaling $1.45 million.

Despite the intentions of Brazil and other nations to work on conservation efforts, global economic demand continues to result in cutover Amazon forests.

Peruvian natives set up a roadblock at the entrance of their town. They are demonstrating against their president's plans to ease restrictions on mining, oil drilling, wood harvesting, and farming in Peru's northern Amazon Rain Forest region. *(AFP/Stringer/Getty Images)*

Change has not been easy. In 2007, the government reported annual forest losses of 11,745 square miles (30,420 square kilometers), but an estimate for 2009 found lowered losses of 3,860 square miles (9,997 square kilometers).

Brazil's population soared to 188 million in 2008. As the fifth-largest nation in the world, its economy has expanded considerably since the mid-twentieth century. Up to 47 percent of the entire Amazon is under direct pressure from human resource consumption from the eight countries that contain it. The largest exported products—soy, timber, and beef, all of which rely on converting forests to farms—have increased steadily since the 1980s.

While protection efforts have increased, they have been faced with the challenge of more aggressive logging companies with more advanced hydraulic machinery able to remove tropical trees faster. Additionally, farms that have been installed in cutover forests have become even more lucrative as they grow genetically modified, heat-resistant varieties of soybeans developed since 1997 to deliver larger crop yields from smaller plots.

Of the products produced from the country's new cropland, Europe consumes 49 percent of the exported soy products, the United States uses 50 percent of the wood, and Russia buys 19 percent of the beef. Multinational companies, such as ADM, Cargill, and John Deere, sow profits from the emergent farm industries.

Another developing issue for Amazonia is the mining of oil and gas reserves by the Brazilian state-owned oil company, Petrobras. A 400-mile (644-kilometer), $1.15 billion dollar gas pipeline is under construction in the state of Amazonas. The expected harmful impacts from deforestation and oil and gas leaks have led to protests.

The sheer size of the Amazon forest and limited staff resources have led the Brazilian government to halt illegal activities by bulldozing nonpermitted roads and exploding large craters in illegal airplane runways in remote locations. Up to 80 percent of illegally deforested areas is located within 18 miles (29 kilometers) of an official government road. By targeting such areas, the government hopes to control future illegal forest removals.

New technology, such as satellite imaging, allows enforcement agents to track illegal logging. These high-resolution images show that most illegal deforestation abruptly ends where indigenous-owned Amazon lands begin. Numerous violent encounters resulting in the deaths of several loggers demonstrate that native peoples are vehement about protecting their lands. A partnership since 1996 between the Amazon Conservation Team—a nonprofit group based in Virginia whose mission is to conserve the biodiversity, health, and cultures of Amazonia—and twenty-eight indigenous tribes has resulted in 40 million acres (16.2 million hectares) of forest being mapped in Brazil, Columbia, and Suriname to enhance local protection efforts.

Brazilian indigenous rain forest tribes, such as the Arara, have sought to use the courts to protect their historic homelands. In 2010, a tribal collaborative, led by the Arara people, tried to halt the construction of the proposed Belo Monte Dam, which would be the third largest dam in the world, in Brazil's state of Para in the heart of the Amazon forest. A Para judge briefly halted the project, citing the "irreparable harm" that it would cause to native residents. However, a few days later, a higher court judge in the Brazilian capital, Brasilia, overturned that ruling, stating that stopping the project would cause devastating economic damage, construction was not imminent, and reviews of environmental impact still were pending.

If the Belo Monte project is built, a 60-mile (96.6-kilometer) stretch of the Xingu River will dry up, 160 square miles (414 square kilometers) of forest will be flooded by a new reservoir, and at least 20,000 residents will be displaced. Criticism of the project includes statements that the facility will only be able to operate at 39 percent of its capacity due to available water flow rates. Brazil currently relies on hydroelectric power for more than 80 percent of its electricity.

Selected Web Sites

Amazon Conservation Team: http://www.amazonteam.org.
Indigenous Dslala Tribe: http://www.smithsonianmag.com/travel/10013581.html.
Smithsonian National Zoological Park, Amazonia: http://nationalzoo.si.edu/
    Animals/Amazonia/.
United Nations Educational, Scientific and Cultural Organization, World Heritage
    Sites: http://whc.unesco.org/en/list/.
WWF–World Wide Fund for Nature (global site): http://www.panda.org.

Further Reading

Bates, Henry Walter. *In the Heart of the Amazon.* 1863. New York: Penguin, 2007.
Bierregaard, Richard, ed. *Lessons from Amazonia.* New Haven, CT: Yale University Press, 2001.
Carson, Walter. *Tropical Forest Community Ecology.* Hoboken, NJ: Wiley-Blackwell, 2008.
Gay, Kathlyn. *Rainforests of the World.* Santa Barbara, CA: ABC-CLIO, 2001.
Goulding, Michael. *Atlas of the Amazon.* Washington, DC: Smithsonian, 2003.
London, Mark. *The Last Forest: The Amazon in the Age of Globalization.* New York: Random House, 2007.
Mock, Greg, ed. *Human Pressure on the Brazilian Amazon Forests.* Washington, DC: World Resources Institute, 2006.
Raffaele, Paul. "Out of Time," *Smithsonian,* April 2005.
Wallace, Scott. "Last of the Amazon," *National Geographic Magazine,* January 2007.

4 Sequoia National Park and National Forest California

The biggest trees on the Earth dominate the landscape they stand over. Towering at heights of up to 310 feet (95 meters), the giant sequoia (*Sequoiadendron giganteum*) can grow up to a remarkable 35 feet (10.7 meters) in diameter. Located in the western United States in central California, these prehistoric-looking trees thrive in the evergreen montane forest (a biogeographic area along mountain highlands just below a subalpine zone) at high altitudes.

This biome features deep snow, cold winters, and short, mild summers. Temperatures range from -10 to 80 degrees Fahrenheit (-23 to 27 degrees Celsius), and rainfall levels are around 90 inches (229 centimeters) per year in the wetter valleys. Two locations, the Sequoia National Park and the Sequoia National Forest, both located in California's Sierra Nevada Mountains, hold most of the existing giant sequoias, which are estimated to span 35,607 acres (14,410 hectares).

The Sequoia National Forest, with thirty-nine sequoia groves (defined as a group of trees), is located near the city of Porterville. At 2,207 square miles (5,716 square kilometers), the Sequoia National Forest ranges in elevation from 1,000 feet (305 meters) to more than 12,000 feet (3,658 meters). Its sequoia groves can be found at elevations between 5,000 feet (1,524 meters) and 7,500 feet (2,286 meters).

Sequoia National Park is located north of the forest near the town of Lemon Cove. It spans 631 square miles (1,634 square kilometers) and has thirty-six groves of sequoias. Both the forest and the park feature a mountainous landscape, with high alpine meadows, steep canyons, river valleys, and rocky bluffs.

Giant sequoia trees are evergreens belonging to the cyprus family. They are among the oldest living individual tree species, some exceeding 2,300 years in age. They weigh in excess of 600 tons (1.2 million pounds, or 0.5 million kilograms).

The conifer leaves are a spiral awl–shape consisting of .25-inch-long (.64-centimeter-long) needles. Sequoias have a reddish-brown bark that is fibrous and up to 2 feet (0.6 meter) thick at the base, which protects the tree against animals, insects, and fires. The trees produce pinecones that are 3 inches (7.6 centimeters) long. These seed-bearing cones have a tough fibrous exterior that does not open until they are approximately twenty years old; opening is induced either by a fire or a widespread insect infestation.

Surprisingly, the main sequoia root system extends down only 6 feet (1.8 meters) into the soil. The tree achieves considerable lateral stability by sending out smaller surface roots up to 300 feet (91 meters) in every direction.

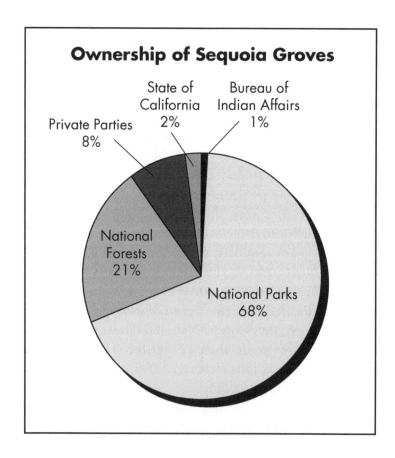

The giant Sequoia (*Sequoiadendron giganteum*) is the biggest tree on Earth. To appreciate the scale of these towering evergreens, note the man standing (bottom right) among this tree's roots. (© Meinolf Zavelberg/Fotolia)

The world's largest living tree, by volume, is located inside Sequoia National Park. This giant sequoia is named General Sherman, after William Sherman, an American Civil War leader. Estimated to be 2,300 years old, the tree is 274 feet (84 meters) tall—as tall as an average twenty-five-story building—and measures 36 feet (11 meters) in diameter at its base. The tree's total volume exceeds 52,508 cubic feet (1,487 cubic meters).

In 2006, one of General Sherman's upper branches fell off during a storm, probably due to wind stress and natural aging. At 100 feet (31 meters) long and 6 feet (1.8 meters) in diameter, this limb, which was larger than most living trees, demolished several small buildings at the tree's base. The world's tallest tree (379 feet/116 meters) is a coast redwood (*Sequoia sempervirens*), located in nearby Redwood National Park. Named "Hyperion," the location of this tree is kept a secret in order to limit the disturbance that would occur if the public could access it.

The geology of central California has slowly created an environment that is ideal for such large trees. Twenty million years ago, California was mostly flat. Considerable mountain uplift occurred when the Pacific Plate (the tectonic plate beneath the Pacific Ocean) collided with the North American Plate. This resulted in the formation of several mountain ranges, including the Sierra Nevada Mountains. The highest mountain in the continental United States, Mount Whitney (14,494 feet, or 4,418 meters), rose within this range.

The groves of giant sequoia trees that at the time covered southwestern Nevada gradually began to spread to the west; they thrived in the newly formed California mountains, where the climate was wetter and cooler. The large trees established themselves in the steep terrain on the mountains' western slopes.

They did not spread to other regions due to drier climates and soil variability, but their close cousins, the redwoods, can be found today throughout the warmer lands of western North America. Other species of sequoia trees, such as the dawn redwood (*Metasequoia glptostroboides*), also exist in southwestern China but only reach heights of around 115 feet (35 meters).

The high-elevation forest community that grows around the sequoias supports a diversity of plants and animals, including one-quarter of the species found in California. The larger montane trees that live near the sequoias include the lodgepole pine (*Pinus contorta*) and white fir (*Abies concolor*), while the understory is filled with smaller trees such as the Pacific dogwood (*Cornus nuttallii*) and the beaked hazelnut (*Corylus cornuta*).

On the ground, a thick layer of pine needles provides extra mulch to protect the soil from the intense mountain sunlight. Ground plants include the native western azalea (*Rhododendron occidentale*), which has large white or pink flowers in late spring. Although it can reach 15 feet (4.6 meters) in height, the azalea usually grows from 4 to 9 feet (1.2 to 2.7 meters) tall.

Animals living in the Sierra Nevada Mountains rely on the forest for shelter, food, and water. Up to 205 species of birds have been recorded in the region. The golden eagle (*Aquila chrysaetos*) has a wingspan of 7 feet (2.1 meters) and a 3-foot-long (0.9-meter-long) body. With dark brown plumage, it blends easily into the colors of the sequoia forest and nests for years in the same trees, up to 250 feet off the ground. From here, these predators with binocular-like eyes (20/200 vision) peer down and dive to snatch up a squirrel, hare, or bird.

The Douglas squirrel (*Tamiasciurus douglasi*) picks or gathers dozens of pine-cones each day. These squirrels have been known to cut as many as 538 green cones from a tree in thirty minutes; they stash the cones in treetop nests to eat the pine nuts at a future time.

Other foraging animals browse along the forest floor, including the mule deer (*Odocoileus hemionus*) and American black bear (*Ursus americanus*). Forest-dwelling predators include the opportunistic wolverine (*Gulo gulo*), a large, ferocious weasel with long claws and razor-sharp teeth.

Human Uses

Native Americans have lived in the central California region as far back as 8,000 years ago. Several groups, related to the Utes and Aztecs of the Southwest, inhabited the foothills of the Sierra Nevada Mountains in the winter months and then moved up to higher terrain for the summer. For example, seminomadic hunter-gatherers such as the Monache people lived in small bands that would camp in a region for a month or two at a time to hunt rabbit, deer, and beaver.

Due to trapping and destruction of continuous habitat, the wolverine (*Gulo gulo*), which once ranged as far south as the midwestern United States, now can be found only in northern parts of North America (as well as in Europe and Asia). An opportunistic scavenger and aggressive predator, a wolverine may challenge animals as much as five times its size, such as black bears (*Ursus americanus*). (© SWP/Fotolia)

River access allowed abundant salmon fishing. The Monache relied on forest resources such as carbohydrate-rich acorns and high-protein pinion nuts.

Although Spanish explorers and missionaries passed through the rugged terrain of the Sierra Nevada Mountains in the eighteenth century, they did not write home about the giant trees or describe them in detail during their travels. Thus, many of the sequoia forests remained "undiscovered" until the 1800s, when logging crews from the eastern United States found the stands and marveled at the trees' height and width.

Initially, loggers noted that parts of the sequoias were brittle and dry and had low tensile strength. They consequently adjusted their cuts to use the outer sections for wood products and the inner sections for large boards and poles.

In the nineteenth century, small towns populated by easterners formed in the mountain region. Around 1849, many of the migrants were gold miners or prospectors for other metals such as copper, silver, and tin.

When the gold rush dried up around 1860, many settlers turned to the forest to log trees, run farms, and raise cattle, providing meat for the expanding California population. Today, small communities still dot the foothill region surrounding both Sequoia National Park and Sequoia National Forest.

Pollution and Damage

Exploreres and settlers brought with them a range of deadly diseases that killed many of the Monache people in the early 1800s. The surviving members of the tribe moved elsewhere in the West.

Environmental damage began in the 1860s, when sequoia trees provided the raw materials for commercial buildings, homes, and farms. Cutting a tree with a diameter of 35 feet (10.7 meters) took ten men more than a week; they sawed smaller, manageable sections to be drawn out by horses. Later, steam-driven tractors and gas-powered bulldozers were used. Each tree usually provided enough board feet to frame forty small houses, shingle the roofs of eighty houses, or build a farm fence with 3,000 posts.

By 1900, up to 400 square miles (1,036 square kilometers) of original sequoia forest were cut down by loggers. Into the mid-twentieth century, entire groves were harvested in the southern Sierras. Many tree parts, however, were not harvested, as they were deemed too brittle. Large piles of bark, dried wood, and treetops were left to rot in the forest. In 1952, a report to the California legislature by the state forester, *Status of Sequoia Gigantea in the Sierra Nevada*, estimated that 34 percent of its sequoias had been decimated.

The U.S. Forest Service, overseeing 2,200 square miles (5,698 square kilometers) of sequoia forest, adopted policies in the 1920s to control forest fires. While they succeeded in protecting private buildings and property, fire control proved detrimental to these forests by altering their natural cycles of reproduction. The heat of an intense fire forces the sequoias' and other trees' pinecones to open and spreads their seeds into the rich ash where a new generation of fast-growing trees is established. By the 1990s, the Forest Service's management techniques had resulted in a dense mass of overgrown forest dominated by Ponderosa pine (*Pinus ponderosa*) and ready to burn. Species such as the sequoia also were shrinking in population from not being able to reproduce due to competition from the faster-growing tree species.

Mitigation and Management

The loss of significant stands of giant sequoias propelled a spectrum of community members to work with state and federal governments to protect the remaining large trees at the end of the nineteenth century. Conservationist John Muir teamed up with George Stewart, the publisher of the Visalia, California, newspaper, and the California Academy of Sciences to lobby the U.S. Congress to purchase and protect the uncut sequoia forests.

In 1890, Sequoia National Park was created as the second national park in the United States. Over the years, the park grew to 404,000 acres (163,493 hectares). In 1978, an adjacent area known as Mineral King was slated to be developed into a ski resort by the Walt Disney Company, but the federal government acquired the land and protected it, making it part of the national park.

Soon after the creation of Sequoia National Park, the Sequoia National Forest was formed in 1908. The mission of the park is preservation of plant and animal life and habitats, while the mission of the forest focuses on public utilization. An overarching policy adopted in the 1950s encouraged the concept of multiple use. These uses included lumber removal, mining, livestock grazing, camping, and the operation of recreational vehicles in the area.

Since the designation of the forest, extensive logging has occurred except in about three dozen old-growth sequoia groves. This has impacted some of the younger sequoia stands that were interspersed in the cutover forest. In 1993, Congress passed the Giant Sequoia Preservation Act, which prohibits any logging within 1,000 feet of a sequoia tree.

Also in response to threats against the giant trees, the federal government created the Giant Sequoia National Monument in 2000 in the heart of the Sequoia National Forest to enhance the level of protection for up to thirty-four old-growth sequoia stands. The monument, overseen by the U.S. Forest Service,

JOHN MUIR ON SEQUOIA TREES

"Any fool can destroy trees. They cannot defend themselves or run away. And few destroyers of trees ever plant any; nor can planting avail much toward restoring our grand aboriginal giants. It took more than three thousand years to make some of the oldest of the Sequoias, trees that are still standing in perfect strength and beauty, waving and singing in the mighty forests of the Sierra. Through all the eventful centuries since Christ's time, and long before that, God has cared for these trees, saved them from drought, disease, avalanches, and a thousand storms; but he cannot save them from sawmills and fools; this is left to the American people."

Source: Sierra Club Bulletin, January 1920.

encompasses 327,769 acres (132,643 hectares) and is divided into two sections, northern and southern. Pressed by logging industry leaders, in 2006 the Forest Service agreed to the removal of up to 7.5 million board feet of non-sequoia stands per year within the monument boundaries. However, that same year a federal judge blocked the Forest Service's decision before work could begin, calling the management plan "decidedly incomprehensible."

In 2003, the Forest Service drafted a Giant Sequoia National Monument Management Plan to manage the future use and protection of the sequoia stands. One element from the plan is the recognition that fire plays a critical role in the natural cycles of the sequoia forest. While a very large fire can burn an entire landscape to the ground, a smaller fire consumes smaller trees and brush, cycling the nutrients back into the soil and allowing sequoias, influenced by the rising heat from down below, to drop their seeds into cleared ash soil.

With the impact from climate change has come more fires and warmer weather. Such environmental changes appear to be positive influences on the ability of this giant tree to expand its presence in California and become reestablished as a long-living, dominant plant species across the greater western United States.

Selected Web Sites

Indian Tribes of Sequoia National Park Region: http://www.nps.gov/history/history/online_books/berkeley/steward2/stewardt.htm.
Sequoia and Kings Canyon National Parks: http://www.nps.gov/seki/; http://www.sequoia.national-park.com.
Sequoia National Forest: http://www.fs.fed.us/r5/sequoia/.
Sequoia Natural History Association: http://www.sequoiahistory.org.

Further Reading

Hartesveldt, Richard. *The Giant Sequoia of the Sierra Nevada*. 1975. Washington, DC: U.S. Department of the Interior, National Park Service, 2005.
Page, David. *Yosemite and the Southern Sierra Nevada*. Woodstock, VT: Countryman, 2008.
Preston, Richard. *The Wild Trees*. New York: Random House, 2008.
Robinson, George. *Sequoia and King's Canyon*. Watertown, MA: Sierra, 2006.
U.S. Forest Service. *Giant Sequoia National Monument Management Plan*. Washington, DC: U.S. Department of Agriculture, U.S. Forest Service, 2003.
Vermaas, Lori. *Sequoia*. Washington, DC: Smithsonian, 2003.
Wadsworth, Ginger. *Giant Sequoia Trees*. Minneapolis, MN: Lerner, 1995.

5 Taiga Forest Russia

Russia's northern hemisphere contains an expansive subarctic region called Siberia. In Russian Turkic, the language of the Tatar people, this word translates into "sleeping land," due to the region's geographic isolation and frigid climate. Across this landscape is a unique forest called the taiga, meaning "marshy forest" in Russian, a name derived from the area's poorly drained soil. This biome is possibly the largest uninterrupted section of woodland on Earth, accounting for one-third of all planet-wide conifer forests.

The region commonly is blanketed with cold air from the Arctic north, resulting in subfreezing temperatures for up to half of the year. Winter temperatures range from -65 degrees Fahrenheit (-54 degrees Celsius) to 30 degrees Fahrenheit (-1 degree Celsius), while summers range from 20 degrees Fahrenheit (-7 degrees Celsius) to 70 degrees Fahrenheit (21 degrees Celsius). The subarctic climate has two primary seasons, a six-month winter and a three-month summer, which makes life in the taiga challenging. Spring and fall each last six weeks at the most.

The taiga stretches 3,540 miles (5,696 kilometers) from west to east and 620 miles (998 kilometers) from north to south, or 2.1 million square miles (5.4 million square kilometers), which is larger than the continental United States. Most precipitation—40 inches (102 centimeters) per year—falls in the form of snow. Siberia's taiga is bordered on the north by the Arctic Circle (66° latitude) and on the south by temperate deciduous forests and grassy plains (50° latitude).

The taiga belongs to a class of forests known as boreal—from Boreas, the Greek god of the north winds.

The soils of the taiga forest are important to understanding the presence of certain plants and trees. The broad majority of the taiga soil is classified as podsol, meaning "under ash," due to a prominent silver-gray-colored middle layer. At the surface, there are 5 inches (13 centimeters) of acidic, dark organic topsoil. The next subsurface layer, which is a quartz-like color, features a high concentration of clay. This poorly drained layer consists of the deposition of glaciers, as well as the seepage of elements such as aluminum from the upper soil. The third layer down from the surface is a red, iron-rich, dense soil mixed

Recent genetic testing has revealed that the endangered Siberian tiger (*Panthera tigris altaica*) is almost identical to the Caspian tiger (*Panthera tigris virgata*) a now-extinct, large cat that once lived in south central Russia and central Asia. Based on this discovery, Russia has offered a trade with Iran aimed at reintroducing the Caspian tiger in Iran and the Asiatic cheetah (*Acinonyx jubatus venaticus*), a critically endangered species, in Russia. (© Pyshnyy Maxim/Fotalia)

with clay and sand. The low-nutrient podsol supports conifer trees and more than 1,000 species of vascular plants, but it is not productive for agricultural use, except to grow hardy crops such as potatoes, carrots, and beets.

The region contains an estimated 1 million lakes, hundreds of thousands of wetlands, and 63,000 rivers, supporting up to 2,300 plant and tree species. The taiga is dominated by fir, larch, pine, and spruce trees; however, Taiga conifers rarely live past 200 years due to pest infestation, fungus growth, and the harsh climate.

The Siberian fir (*Abies sibirica*) is the most common taiga tree species. It is a softwood evergreen, growing up to 115 feet (35 meters) high and 3 feet (0.9 meter) wide. This native species is able to survive in poorly drained soils where low light and subzero temperatures are the norm. The Siberian larch (*Larix sibirica*) repels the cold weather with thick bark and a frost-resistant vascular system, which reduces wood rot. The Siberian spruce (*Picea obovata*) grows up to 100 feet (30 meters) and has distinctive 4-inch-long (10-centimeter-long) pine-cones with tough, weatherproof scales.

In Siberia, fires are common during the summer and fall. Spurred on by the highly resinous sap of the dominant conifer trees, large wildfires spread quickly. The fires remove the lower vegetation in choked and dense areas and trigger natural reseeding from heat-opened pinecones.

The taiga conifers often form a tight canopy, which allows for little direct light penetration to the forest floor. Understory shrubs and bush species must be tolerant of low-light and cold conditions in order to survive.

Especially important to mammals and birds are the bushes that produce berries. The cloudberry (*Rubus chamaemorus*) produces a bright orange, raspberry-sized fruit rich in carbohydrates and vitamin C. Called *moroshka* in Russian, these bushes are common near wetlands.

The low-bush mountain cranberry (*Vaccinium vitis-idaea*) is 15 inches (38 centimeters) high; even though this variety technically is deciduous, the plants stay foliated year-round. The cranberry's red berries provide an excellent food source rich in sugars and antioxidants for dozens of species of foraging mammals.

The taiga forest supports an estimated 203 bird and eighty mammal species. The red crossbill (*Loxia curvirostra*) is a finch-like bird that lives in dense firs, spruces, and hemlock trees. This bird uses its powerful hooked beak to pry open pinecone scales, and its long tongue pulls out the protein and fat-rich pine nuts.

With large sections of forest uninhabited by humans, Siberian wolves (*Canis lupus communis*) thrive, with up to one individual per square mile (up to one per 2.5 square kilometers). They hunt Eurasian elk (*Alces alces*) in packs of up to ten wolves.

Another predator, inhabiting the far eastern taiga, is the Siberian tiger (*Panthera tigris altaica*), the world's largest cat species. At up to 450 pounds (204 kilograms), this apex feline has a thick coat and insulated paws to protect itself in deep snow. Its diet consists mainly of boar, elk, and reindeer. The Siberian tiger was considered extinct from western and central Asia by 1920. At one point, extensive poaching between 1910 and 1975 across the eastern forest had reduced the population to less than 250 of these big cats, but recent conservation efforts have helped repopulate the central Siberian region, bringing the count up to about 480 tigers.

Human Uses

Russia's Siberian taiga occupies nearly 70 percent of the country but holds less than 28 percent of the population. With 20 million people spread across ten states, from the Baltic Sea in the west to the Sea of Japan in the east, the density of humans in the forest is 1.15 persons per square mile.

The taiga is anchored by two cities: St. Petersburg in the west and Yakutsk in the east. St. Petersburg had a population 4.6 million in 2006. It was the Russian capital for two centuries, from 1713 until 1918. The city was known as Leningrad from 1924 to 1991. The fourth-largest city in Europe, St. Petersburg is situated on the shores of the Baltic Sea and at the edge of the taiga. The city of Yakutsk is located 5,700 miles (9,171 kilometers) to the east. It had a population of 245,600 in 2009. Founded in 1632 as a fishing port on the Lena River, Yakutsk is the largest city built almost entirely on permafrost.

Both St. Petersburg and Yakutsk rely on a range of industries, including financial companies and other corporations, mining, logging, farming, industrial processing, such as smelters, and factories. Scattered across the taiga are hundreds of small communities, many rural and isolated.

Approximately 180,000 people, made up of thirty indigenous groups, including the Evenki People, live in the taiga and rely on the forest. Some of these groups migrate seasonally, moving in the summer onto the marshy arctic plains and returning in winter to the dense coniferous forests. For instance, the Nenet people, with a population of about 10,000, are reindeer herders who process most of their meat and clothing from this one species (*Rangifer tarandus*).

Historically, many taiga native peoples have not fared well under Russian rule. Groups faced colonial oppression, including forced relocation, banning of traditional spiritual and social ways, removal of their young to foster homes, settlement on collective farms, and forced education and employment.

Surviving indigenous groups try to balance their modern lifestyle with traditional values. Many in the taiga have grown reliant on Russian commerce and transportation systems. Some groups, however, have tried to remain isolated

Reindeer (*Rangifer tarandus*) provide the Nenet people with food, clothing, shelter, and transportation. (*Maria Stenzel/National Geographic/Getty Images*)

from the mainstream Russian population. The environmental impacts of mining, logging, and oil and gas extraction continue to oppress these peoples, but their seminomadic ways provide them with resiliency, especially in the isolated portions of the forest.

Pollution and Damage

Russia is 6.5 million square miles (17 million square kilometers) in size, making it the largest country in the world. It has more than 140 million residents who rely heavily on its natural resources. Russia's massive taiga forest contains extensive stands of trees, considerable oil and gas reserves, and widespread mineral deposits. Given the historic political and economic difficulties of the country, regulation and protection of the forest has been a low priority. With most of the taiga under either local or state control, estimates for illegal logging during the late twentieth century have approached 30 percent of the total trees removed.

Intense logging in the taiga began in the 1950s as Russia (then the Union of Soviet Socialist Republics, or Soviet Union) positioned itself as a world superpower. Forests were cut down to help build government infrastructure, and oil was drilled to provide fuel. Up to 98 percent of the tree harvests were performed by clear-cutting of large areas, rather than removals of smaller sections of forest. Logging between 1992 and 2002 totaled 7 billion cubic feet (0.2 billion cubic meters) each year, generating a total of $461 billion. The logging community employed up to 18 percent of the population living in the taiga during this period.

Another factor in the survival of the taiga forest is the expansion of factories in the region. Aluminum, gold, lead, nickel, and other metal processing facilities located across Siberia emit toxic pollution, such as nitrogen oxides

EVENKI PEOPLE

Region of Taiga: Western Sibera, near the Ob River
First Encounter with Russians: 1606
Population in 1897: 62,068
Population in 2002: 35,527
Autonomous Settlement Area: Krasnoyarsk, 296,526 square miles (768,002 square kilometers); the Evenki also live in five other Russian regions and in China
Living Patterns: Seminomadic
Food Sources: Sheep, reindeer, fish, basic crops
Dwellings (historic): Conical tents
Dwellings (current): Wooden buildings
Religion: Shamanism, Christianity (twentieth century)
Written Language (Evenki): Created in 1920

Source: The Red Book of the Peoples of the Russian Empire, 2001.

and sulfur dioxide, into the atmosphere. The result has been widespread water pollution and acid rain across the forest.

The central Siberian city of Norilsk was established in the 1920s. Its population of 300,000 specializes in industrial processing; the city is the world's largest producer of cobalt, copper, and nickel. In 2000, the combined annual industrial discharge from Norilsk exceeded 2 million tons of sulfur dioxide—forty times the limit recommended by the World Health Organization—19,000 tons of nitrogen oxides and 10,000 tons of carbon monoxide. Norilsk annually discharges up to 14 trillion gallons (53 trillion liters) of liquid waste, often laced with heavy metals, PCBs (polychlorinated biphenyls, an organic compound used in industrial manufacturing), and toxic chemicals, into local waterways. Up to 1,158 square miles (2,999 square kilometers) of taiga forest located downwind of Norilsk have been heavily impacted; there is widespread evidence of the effects of acid rain for hundreds of miles.

At the same time, Russia's oil and gas industry has profited financially from deposits in the taiga. Neighboring industrial countries, such as China, India, Japan, South Korea, and Taiwan, all have pursued substantial contracts either to purchase or to develop oil and gas resources in the region. China, poised to be the second-largest oil consumer in the world, after the United States, has been the most aggressive, building a 3,055-mile (4,915-kilometer) natural gas pipeline from eastern Siberia to northeast China at a cost of $17 billion.

The construction of the Trans-Siberian Railway in 1903 spurred widespread economic development in the taiga region. Spanning 5,770 miles (9,284 kilometers) from St. Petersburg to Lake Baikal, the railway crosses eight time zones. In 2004, the Russian government announced a goal to complete the long-planned, 6,600-mile (10,619-kilometer) Trans-Siberian Highway from St. Petersburg to Vladivostok, at the Sea of Japan.

Limited environmental oversight of the highway construction has threatened virgin taiga tree stands. Insufficient funding has left sections (up to 150 miles, or 241 kilometers, long) as basic gravel roads. Despite these and other challenges, sections of the highway project have opened access to the taiga, brought economic prosperity to many communities, and facilitated population growth.

Mitigation and Management

Russia's taiga composes up to one-fifth of all forestlands on Earth, but efforts to protect the region have been challenging. The thirst for short-term economic gain has emphasized logging, mining, and drilling. Spilled oil has contaminated soils and waterways, forcing indigenous people to relocate from traditional hunting grounds. Damage from clear-cuts has increased surface temperatures, reduced water storage capacity, decreased animal and plant populations, and reduced oxygen generation and carbon storage. Widespread acid rain has resulted in dead, stunted, defoliated, and dying trees.

The Russian federal government began to protect forest regions in 1988 with conservation laws for both public and private forests. Federal, state, and local parks have grown since the mid-1980s to include fifty mountain parks, thirty-eight forest nature reserves (known as *zapovedniks*), and two dozen national parks. A 1993 federal law, the New Forestry Act, set joint forestry management goals for the national and local authorities.

A 1994 regulation limited industrial clear-cuts to 123 acres (50 hectares) in size. In 2007, the federally protected forests—where clear-cutting is strictly limited—totaled 938,000 square miles (2,429,420 square kilometers), about the size of the states of Texas and Alaska combined.

FOREST CODE OF THE RUSSIAN FEDERATION

Article 1: Key Principles

1. Sustainable forest management, biological diversity, conservation in forests, and enhancement of this potential.

2. Maintenance of habitat-forming, water conservation, protection, sanitation, recreation and other beneficial functions of forests, to ensure that each person could exercise the right for a healthy environment.

3. Use of forests with due regard to their global environmental significance, as well as taking into account the length of their cultivation and other natural properties.

4. Multiple-purpose, sound, continuous, non-depleting use of forests to satisfy society's needs for forest and forest resources.

5. Renewal of forests, improvement of their quality and yield.

6. Ensured protection of forests.

Source: Russian Federal Forestry Agency, 2007.

PROTECTED TAIGA FOREST RESERVES

Name	Year Created	Geographic Location	Size (Square Miles)
Baikal State Nature Reserve	1979	Eastern	453
Pechoro Reserve	1930	Western	2,785
Pinezh Reserve	1975	Western	159
Sayano-Shushen Reserve	1976	Central	1,504
Sokhondo Reserve	1979	Eastern	815
Total Square Miles:			5,716

Source: Taiga Rescue Network, 2005.

International financial assistance, through direct investments or loans, has played a key role in protecting the Russian forest landscape. The Trans-Siberian Highway was built with funds from the European Union and The World Bank, and both of these institutions make it a policy of requiring environmental protection efforts. This has resulted in efforts to reduce surface runoff and preserve fragile wetlands.

The Russian government is an active member of the United Nation's Forum on Forests. The principal law guiding how forests are managed in Russia, the Forest Code of the Russian Federation, was rewritten in 2007 to emphasize modern forestry management practices; however, some critics say that the code does not provide needed protections for the indigenous peoples who live in and rely on the forests.

The taiga forest's daunting size challenges modern regulators in their attempts to protect the forest land. Despite better forest management practices, illegal logging represented 25 percent of the trees removed in 2007. As of 2009, weak local enforcement of forest protection laws continued to affect the lives of many rural native people.

Groups such as the indigenous Udegei people from the eastern taiga have responded directly to illegal logging on their land by using armed patrols to defend the forest against such encroachment. In 1992, a South Korean logging venture, approved by the local forest administrator, was stopped when the tribe used hunters with rifles to guard access points. In 2001, a French timber company leased a 1.7 million acre (0.7 million hectare) parcel of Udegei-owned forest from the local government. The Udegei people took the case to court and won a reprieve when the Russian Supreme Court determined that their land had been illegally utilized.

Selected Web Sites

Global Forest Watch: http://www.globalforestwatch.org.

Russian Federal Forestry Agency: http://www.rosleshoz.gov.ru/english.

Taiga Rescue Network: http://www.taigarescue.org.

Wildlife Conservation Society Siberian Tiger Project: www.wcs.org/international/asia/russia/siberiantigerproject.

Further Reading

Day, Trevor. *Taiga*. New York: Chelsea House, 2006.

Hays, Forbes. *Taiga*. Washington, DC: American University, Trade Environment Database, 1998.

Johansson, Philip. *The Forested Taiga: A Web of Life*. Berkeley Heights, NJ: Enslow, 2004.

Kasischke, Eric, ed. *Fire, Climate Changes, and Carbon Cycling in the Boreal Forest*. New York: Springer, 2000.

Kolga, Margus. *The Red Book of the Peoples of the Russian Empire*. Tallinn, Estonia: NGO Red Book, 2001.

Scherer-Lorenzen, Michael, ed. *Forest Diversity and Function: Temperate and Boreal Systems*. New York: Springer, 2005.

Shugart, Herman, ed. *A Systems Analysis of the Global Boreal Forest*. New York: Cambridge University Press, 2005.

6 Rain Forests of Costa Rica

Costa Rica spans only one-tenth of 1 percent of the Earth's landmass, but the country contains diverse forests that hold 5 percent of planet-wide terrestrial species. This small Central American country was named "rich coast" in Spanish by Christopher Columbus when he received gifts of gold after landing there in 1502.

Costa Rica is bordered on the west by the Pacific Ocean and on the east by the Caribbean Sea. Covering 19,730 square miles (51,101 square kilometers), the landscape is a mix of lowlands, hills, and mountains. It also hosts active volcanoes, part of the Andean–Sierra Madre mountain chain.

The country sits between 8° and 11° latitude north of the equator, with the nations of Panama to the south and Nicaragua to the north. The second smallest country in Central America, Costa Rica is environmentally important for its varied biomes, including coastal reefs, estuaries, whitewater rivers, diverse wetlands, expansive lush forests, steep mountains, and dry grasslands.

Forest Types

Forests cover roughly 47 percent of Costa Rica, or 9,231 square miles (23,908 square kilometers). There are four types of forest: coastal mangroves, lowland rain forests, dry deciduous forests, and montane rain forests.

The climate changes with the topography. Lower-altitude plains average from 70 to 90 degrees Fahrenheit (21 to 32 degrees Celsius) and receive 82 inches (208 centimeters) of rain per year. In the higher-elevation rain forests, the temperature ranges from 60 to 76 degrees Fahrenheit (16 to 24 degrees Celsius) and is often wet, receiving up to 140 inches (356 centimeters) of rain per year. In this tropical environment, 12,119 species of plants and 1,200 species of trees abound, as well as 838 species of birds, 258 species of reptiles, 232 species of mammals, and 183 species of amphibians.

Coastal Mangroves

With more than 800 miles (1,287 kilometers) of coastline, Costa Rica boasts hundreds of acres of tropical mangrove trees that grow at the edges of rivers where salt water and freshwater meet. Instead of a typical terrestrial upland forest, the mangrove tree lives in a high-saline (up to 90 parts per thousand) wetland habitat where species such as the black mangrove (*Avicennia germinans*), red mangrove (*Rhizophora mangle*), and white mangrove (*Laguncularia racemosa*) grow in or at the edge of the water.

All three of these tree species have evolved over thousands of years with physical adaptations that allow them to thrive in a salty, coastal environment.

Mangroves have spidery roots that are anchored in the soft saturated soils. To cope with high salinity and low oxygen content, both the black and white mangroves send out snorkel-like roots that are exposed to the air at low tide, where oxygen can be absorbed through the air. During high tide, the root pores close, effectively keeping the salt water out. The red mangrove, which can grow up to 80 feet (24 meters) high, can exist in submerged or

The purpose of the spider leg–like roots of the walking palm (*Socratea exorrhiza*) has been debated for some time. Among the hypothesized purposes is the ability for the tree to use new roots to "walk" horizontally toward a patch of sunlight, thus giving this species its name. *(Michael Melford/National Geographic/Getty Images)* upland conditions. The tree drops its seeds, called propagules, as miniature baby trees, which are distributed by water movement.

Lowland Rain Forests

Lowland rain forests are found along the eastern Caribbean side of the country. Upper canopy trees such as the mahogany (*Swietenia macrophylla*) grow 210 feet (64 meters) tall to compete for strong sunshine. Underneath, different canopy layers with varying light levels, temperatures, and nutrient needs support their own microclimates.

Thick, woody vines are ever present, climbing up anything they can wrap themselves around. Various species of epiphyte vines, which lack typical root systems, use a distinct leaf structure to capture falling rain or organic material and absorb nutrients.

Along the ground a struggle for sunlight and space forces plant species to adapt. The walking palm (*Socratea exorrhiza*) grows on top of several feet of spider leg–like roots, continuously growing new roots, which are believed to slowly propel the tree horizontally toward open sunlight.

Dry Deciduous Forests

The northwestern and western sides of Costa Rica contain seasonal dry deciduous forests that make up less than 2 percent of the total forested area. In response to less rain, the trees and plants in this ecosystem drop their leaves—typically between the months of January and March—to conserve water.

The large, green ceiba (*Ceiba trichastandra*) has green bark that can perform photosynthesis in the dry months when its leaves are gone. Other species, such as the rain tree (*Samanea saman*), can close their leaves in a daily process called nyctinasty. Each night, when the temperature drops, the leaves roll up to conserve moisture and reduce energy use.

Montane Rain Forests

Up to half of the geography of Costa Rica sits above 3,400 feet (1,036 meters). This elevated area contains montane rain forests, which are bathed in rain and mist from the intersection of warm tropical and cool mountain air. These "cloud" communities are exposed to lower temperatures than other parts of the Costa Rican landscape.

Due to the altitude and strong winds, most trees that grow in this region are smaller than lowland species. For instance, the guarumo tree (*Cecropia peltata*) rarely exceeds 50 feet (15 meters) in height and has broad leaves to capture the

This "strangler fig," or banyan tree (Ficus benghalensis), begins its life as an epiphyte (air plant) in the upper branches of trees; it sends thin roots downward to the soil and then grows upward to the sunlight. Over time, this member of the Ficus family strangles its host tree, leaving only a shell of what was once there. *(Yuri Cortez/ AFP/Getty Images)*

sun on angled mountain slopes. Guarumo sap is used by indigenous people medicinally to help increase blood circulation.

Across the canopy, a dense wall of greenery grows, including a range of "air plants," also known as epiphytes, whose roots gather nutrients from rain and air instead of from the soil. Spanish moss (*Tillandsia usneoides*) is a common epiphyte, with slender vines and small leaves. Growing in thick layers, the moss intercepts sunlight, eventually causing its host tree to lose leaves and branches from the smothering effect of the vines.

Mountains and Volcanoes

The geology of Costa Rica has undergone many changes due to its placement at the intersection of the Caribbean, Cocos, Nazca, North American, and South American tectonic plates. The tectonic pressure over the last 58 million years has caused considerable mountain uplift, resulting in the formation of the Andean–Sierra Madre Mountains, which cross the country diagonally from northwest to southeast.

Costa Rica's highest mountain is the Chirripó Grande (12,599 feet, or 3,840 meters), which is located in the middle of the Talamanca Range in the southern part of the country. The region surrounding the mountain has been protected from development by its steep, rugged terrain, as well as by the limited local road system.

Numerous volcanoes are present in the northwest, central, and southeast portions of the country. Due to the buildup and release of underground tectonic pressure, earthquakes are common. A 1910 earthquake in the city of Cartago killed more than 700 people; another in 2004 registered a magnitude of 6.4 and took eight lives. In 1989, 16,000 small tremors were measured over a period of sixty days.

The Cordillera de Tilarán range in the south contains the 5,437-foot (1,657-meter) volcano Arenal. This stratovolcano (a tall, conical volcano built through rapid eruption) has small eruptions a few times a year, with major eruptions spaced out every few hundred years. A 1968 eruption killed eighty-seven residents and demolished three villages.

Human Uses

Costa Rica, which gained independence from Spain in 1821, is divided into seven regions and eighty-one counties. It has a modern economy that is heavily reliant on tourism. The majority of the country's population (4.2 million in 2009) is concentrated in the central highlands and the capital of San José. The Spanish-speaking residents are primarily of African, Caribbean, and North and South American ancestry. As many as 42,000 indigenous peoples live in Costa Rica, especially in the far north and south.

The indigenous Bribri people have a close relationship with the forests of Costa Rica. The tribe numbers around 11,500 and inhabits pockets of steep rain forest in the southeast. Living in the Talamanca Mountains, small bands of Bribri farm crops such as beans, corn, and squash. Their religion centers on the animal world and a designated healer, called an *awa*, who utilizes the plants of the rain forest to treat a variety of illnesses. Numerous forest products have medicinal uses.

Due to extensive rain and year-round warmth, up to 10 percent of the Costa Rican landscape is farmed, especially in the central plains. Many years of volcanic deposition have resulted in nutrient-rich soil. The largest crops include bananas, coffee, pineapples, sugar, and tobacco. Up to 15 percent of the population works in agriculture, while 22 percent works in light industry such as domestic manufacturing, cloth making, and food processing.

Widespread conservation areas on public land protect many forests, but logging on private land is common. In 2006, the logging industry employed 7,000 residents and generated $122 million from tropical "roundwood" (wood from trees less than a foot in diameter), much of which is sold as firewood.

Pollution and Damage

Development has steadily increased across Costa Rica. In order to provide more land for human use, up to half of Costa Rica's rain forests have been clear-cut and the land developed since 1940. These clearings provided space for expanding communities and roads. In areas featuring rich volcanic soil, most of the cleared lands were turned into productive farmland. However, many of the deforested parcels around the country were abandoned when conditions such as poor drainage, leading to excessive rainwater runoff, or thin soils were discovered.

Between 1990 and 2005, clear-cutting destroyed 6.7 percent of the total rain forests (29.4 percent of this amount consisted of virgin forests). This represents one of the most devastating cases of deforestation in Central America during that fifteen-year time frame.

In northern Costa Rica, the Tilarán Mountains climb to heights of 6,070 feet (1,850 meters), and these peaks are filled with cloud forests. Niche plant species thrive in the conditions of almost constant mist, rain, and fog, which create a montane biome.

The cloud forest relies on trade winds that blow from the Pacific Ocean across the mountains, from west to east. When warm, moisture-laden air comes into contact with cool mountain air, it condenses to form precipitation. Regional deforestation and an increase in ranching, farms, and development have resulted in an increase in temperatures of the surrounding land. The temperature increase has resulted in more evaporation and less mountain condensation, thereby reducing moisture at these high elevations and threatening the cloud forest species.

Costa Rica imports up to 18 million pounds (8.2 million kilograms) of pesticides annually, mainly for banana and coffee farms. Since the late 1980s, farms have grown in size and have required newer types and greater concentrations of pesticides as the market increasingly has demanded blemish-free products.

Researchers have discovered that when pesticides are applied in lowland farms that tend to have very warm weather (generally averaging above 72 degrees Fahrenheit, or 22 degrees Celsius), a portion of the chemical remains in a gaseous form. It is then carried by winds up the mountain slopes, 15 to 50 miles (24 to 80 kilometers) away, where precipitation then deposits the chemicals into formerly pristine forests. Researchers have concluded that the concentration of such pesticides on Costa Rica's mountain slopes likely has played a significant role in the disappearance of a number of frog and other amphibian species.

RAIN FOREST BIRDS OF COSTA RICA

Common Name	Scientific Name	Characteristics
Emerald Toucanet	*Aulacorhynchus prasinus*	Found in mountainous forests, this bright green bird has a large yellow bill adapted for eating dozens of different kinds of fruit.
Green Honeycreeper	*Chlorophanes spiza*	Occupying the rain forest canopy 50 to 100 feet (15 to 30 meters) off the ground, this small tanager-like bird feeds on flower nectar, fruits, and insects.
Red-footed Plumeleteer	*Chalybura urochrysia*	This large, short-billed hummingbird is territorial. It lives at the edges of forests and feeds on the nectars of flowers found growing in adjacent fields and clearings.
Resplendent Quetzal	*Pharomachrus mocinno*	Living in the cloud montane forest region, these birds have 60-inch-long (152-centimeter-long) tail feathers. They were cherished as "snake gods" by pre-Columbian cultures.
Rufescent Tiger-Heron	*Tigrisoma lineatum*	Living in wetlands and along the banks of waterways, this species hides in undergrowth and captures fish and invertebrates with a long, sharp bill.
Sunbittern	*Eurypyga helias*	Similar to herons, this bittern species hunts for small vertebrates along the forest floor. Its display of unique multi-colored wings wards off predators.

Source: Richard Garrigues. *The Birds of Costa Rica.* Ithaca, NY: Cornell University Press, 2007.

Mitigation and Management

By 1990, Costa Rica's growing economy was heavily impacting its rain forests. More businesses, growing tourist resort communities, and expanding animal and crop pastures resulted in extensive clear-cutting and the burning of private

CARBON NEUTRAL FACTS ABOUT COSTA RICA

- In 2002, the country emitted roughly 5.8 million tons of carbon dioxide from industries, fires, transportation, and other sources.

- Carbon emissions of 5.8 million tons amount to 1.5 tons per person (compared to 5 tons per person in the United States).

- In 2007, 94 percent of electricity came from renewable resources, such as geothermal energy, hydroelectric sources, and wind farms.

- A "C-Neutral" label, available in 2010, will identify businesses and products that eliminate or offset all carbon emissions.

- In 2008, researchers estimated that Costa Rican forests absorb 2.5 million tons of greenhouse gases per year.

- Costa Rica's goal is to achieve carbon-neutral status by 2021.

Source: Costa Rica Ministry of Environment, Energy and Telecommunications, 2008.

forests. Public outcry over the loss of valuable rain forests and input from a range of scientific and conservation organizations eventually brought about government action to improve forest and biodiversity management.

One of the first successes was a government program introduced in 1997 that paid landowners not to log their forests. By 2006, the program was paying $15 million per year to 8,000 landowners, or an average of $1,875 each.

Another significant change was the government's conservation of almost 28 percent of the country, an unprecedented level of protection to species-rich forests. A total of 190 biological reserves, national parks, and wildlife refuges were established throughout the country. Between 1990 and 2006, restoration efforts resulted in 51 percent forest coverage across Costa Rica, a 10 percent increase.

Despite these accomplishments, persistent financial debts, corruption scandals, and intermittent funding for conservation programs have made it difficult for government leaders to maintain and improve forest management efforts. While the framework and regulations to protect wide swathes of forest are in place, troubled government systems and poor enforcement continue to result in illegal logging and the poaching of animals, such as the endangered jaguar (*Felis onca*).

As a result, the Costa Rican government has collaborated with colleges and universities, nonprofit organizations, and other countries to better manage its forests. One 1995 effort with the Nature Conservancy resulted in the creation of the 103,258-acre (41,787-hectare) Corcovado National Park on the Osa Peninsula on the southwest Pacific coast. The project included establishing a new ranger program, including training, vehicles, equipment, and facilities. Up to half of Costa Rica's endangered species lives on this peninsula, including the harpy eagle (*Harpia harpyja*), ocelot (*Leopardus pardalis*), Baird's tapir (*Tapirus baidii*), and giant anteater (*Myrmecophaga tridactyla*).

In 2007, Costa Rica agreed to a unique "debt-for-nature swap" with several partners to enhance the amount of protected forest habitat. Under the terms of the agreement, the United States and two conservation organizations—the Nature Conservancy and Conservation International—will assume $26 million of Costa Rica's debt over a sixteen-year span in exchange for improved management of rain forest areas.

Also in 2007, the Costa Rican government announced an ambitious and complex plan to be the first "carbon-neutral" country by 2021. A new gas tax and a visitors' tax will fund offsets for the nation's carbon dioxide emissions. The plan also includes reducing emissions through modernizing transportation systems, improving farming practices and industry efficiency, and increasing the acreage of rain forests. By 2008, the government had planted up to 5 million new trees to increase carbon absorption capacity.

Selected Web Sites

Corcovado National Park: http://www.corcovado.org.
Costa Rica Parks: http://www.centralamerica.com/cr/parks/.
Costa Rica Tourism and Travel Bureau: http://www.costaricabureau.com.
Costa Rican National Biodiversity Institute: http://www.inbio.ac.cr/en/.
National Public Radio, Carbon-Neutral Costa Rica: http://www.npr.org/templates/story/story.php?storyId=19141333.

Further Reading

Allaby, Michael. *Tropical Forests.* New York: Chelsea House, 2006.
Allen, William. *Green Phoenix: Restoring the Tropical Forests of Guanacaste, Costa Rica.* New York: Oxford University Press, 2001.

Baker, Christopher. *Costa Rica*. New York: DK, 2005.

De Camino, Ronnie. *Costa Rica: Forest Strategy and the Evolution of Land Use*. Washington, DC: World Bank, 2000.

Frankie, Gordon. *Biodiversity Conservation in Costa Rica*. Berkeley: University of California Press, 2004.

Nelson, Andrew. *Protected Area Effectiveness in Reducing Tropical Deforestation*. Washington, DC: The World Bank, 2009.

7 Congo Forest Central Africa

In Central Africa, the Democratic Republic of the Congo (DRC) contains up to 17 percent of the world's tropical rain forests. Bordered by the 2,920-mile (4,698-kilometer) Congo River in the north and the Kasai River in the south, this equatorial forest encompasses 905,365 square miles (2.3 million square kilometers), an area approximately equal to one-third of the entire United States. Nine countries surround the DRC, including Angola, Burundi, Cabinda, Central African Republic, Rwanda, Sudan, Tanzania, Uganda, and Zambia.

DRC rain forests span an area that lies between high-elevation mountain lakes and volcanic peaks on the east and by a narrow section of land only 25 miles (40 kilometers) wide that fronts the Atlantic Ocean on the west. This region contains half of Africa's forested habitat, including a tremendous diversity of living creatures: 10,000 species of woody and vascular plants, 617 species of birds, 450 species of mammals, 350 species of trees, 208 species of amphibians, 122 species of fish, 21,000 species of insects, and 304 species of reptiles.

The Congo River is a central feature of this rich equatorial forest. The river originates among 16,000-foot (4,877-meter) volcanic peaks and flows from east to west, ending at the Atlantic Ocean. The Congo provides an average freshwater flow of 1.4 million cubic feet (40,000 cubic meters) per second. As the second-longest river in Africa (after the Nile River), this essential runoff from the mountains and high-elevation lakes supports a diversity of forest types.

The region's tropical climate experiences two seasons. The rainy season is from March to November, followed by a dry season from December to February. Precipitation ranges from 70 to 150 inches (178 to 381 centimeters) per year, and average temperatures are from 64 degrees Fahrenheit (18 degrees Celsius) in the mountains to 85 degrees Fahrenheit (29 degrees Celsius) near the coast.

The DRC rain forest is divided into two geographic areas: the central basin and the eastern highlands. The heart of the rain forest hosts the central basin, which is divided in half by the equator and makes up 45 percent of the region's forested areas. It contains two types of woodland: eastern swamp forests and central lowland forests.

The swamp forests are periodically flooded in the rainy season, and nutrient-rich sediment is distributed across the flatlands. Once the rains slow, a blanket of grasses, plants, and trees thrive in the rich soils. The woody materials are routinely cleared for firewood and heavily grazed. Flooding forces both humans and many animals to leave the area, thus limiting natural resource overuse and promoting regrowth. The central lowland forest is a rain forest that supports a unique biome called the savanna. Widely spaced trees and grassy plains define this region. The savanna can be found at elevations up to approximately 2,500 feet (762 meters), receives about 48 inches (122 centimeters) of rain per year, and often is cooler than the lowlands.

Trees such as the sausage tree (*Kigelia africana*) and the black plum tree (*Vitex doniana*) thrive in the open canopy environment. The sausage tree grows up to 60 feet (18 meters) tall and produces long, round fruit that is pulpy and high in fiber. The black plum tree is smaller, about 50 feet (15 meters) tall, and it sends shoots underground to grow new trees. Its smaller, 2-inch-long (5-centimeter-long) fruit turns black when ripe and tastes like prunes. Black plum wood is highly termite resistant and thus this tree species lives longer than many others. The fruit from both trees is consumed by a range of mammals, including elephants, giraffes, monkeys, and birds, such as cockatoos and parrots.

The eastern highlands have an elevation of approximately 4,000 feet (1,219 kilometers) and run for 950 miles (1,529 kilometers) north to south along the edge of the Great Rift Valley and 300 miles (483 kilometers) east to west. These highlands encompass about 55 percent of the Congo Rain Forest.

The steeper terrain supports trees such as the ebony tree (*Pithecellobium flexicaule*) and the wenge tree (*Millettia laurentii*). The ebony is an evergreen species that grows to 35 feet (10.7 meters) and can tolerate poor soils. The wenge has a dark gray wood grain and is heavy and dense. When cut, the tree's oils cause a skin reaction similar to poison ivy.

At 6,500 feet (1,981 meters), a montane forest emerges, and, at the foot of the Ruwenzori Mountains, shorter tropical forest stands are interspersed with grassy meadows. The eastern highland's tallest peak, Mount Nyamuragira,

CONGO SAVANNA MAMMALS

Common Name	Scientific Name	Characteristics
African Buffalo	*Syncerus caffer*	Growing to about 2,000 pounds (907 kilograms), this large grazer spends up to ten hours a day browsing in grasslands.
African Elephant	*Loxodonta cyclotis*	The savanna is a common habitat for the biggest of rain forest mammals, due to the wide variety of plants, grasses, and small trees available for its consumption.
Chimpanzee	*Pan troglodytes*	This primate, which lives up to forty years in the wild, can grow up to 90 pounds (41 kilograms) by the age of ten. It makes nests in trees and eats fruit, insects, and, rarely, meat.
Leopard	*Panthera pardus*	This nocturnal big cat often sleeps in trees and hunts for food in open ranges. It can live up to twenty years.
Spotted Hyena	*Crocuta crocuta*	Working in groups of up to forty, these predators roam 25-square-mile (65-square-kilometer) tracts hunting prey, including gazelles, wildebeests, and zebras.

Source: Jonathan Kingdon. *The Kingdon Field Guide to African Mammals*. New York: Academic Press, 2003.

is 10,033 feet (3,058 meters) high, and it is an active volcano, with occasional eruptions of smoke and lava.

The African forest elephant (*Loxodonta cyclotis*) is the largest mammal in the Congo forest. It has a population of approximately 64,000 and lives in both the central basin and the eastern highlands. Measuring 10 feet (3 meters) tall and weighing up to 8,000 pounds (3,629 kilograms), they rely on a vegetarian diet of young tree leaves, grasses, fruits, and pulpy branches.

These majestic and powerful creatures play a central habitat-forming role in the dense rain forest. By grazing up to 300 pounds (136 kilograms) of vegetation in a single day, the forest elephants open up the lower forest, which allows new vegetation and animals to thrive. In addition, to receive sufficient nutrients, the elephants use their powerful trunks to push deep into thick, muddy freshwater pools, causing salts and other needed minerals to bubble up. These nutrients encourage plant and tree growth by providing phosphorus, sulfur, and nitrogen to adjacent soils.

Human Uses

West Africa was one of the primary locations of early human evolution, and the forest played a central role in ancient societies. An early human, *Homo habilis,* lived in this region 1.6 million years ago and engineered some of the earliest wooden and stone tools, used for defense and daily food gathering and processing. More advanced hunting and gathering groups thrived in the lush Congo rain forests 45,000 years ago during the late Stone Age. Trade of goods among communities also brought a greater variety of food sources, seeds, and improved technology, such as fire-starting tools.

The development of agricultural systems, the ability to form metals, including copper, gold, iron, and silver, and a culturally diverse society led to the formation of four separate kingdoms in the Congo basin by 1400 C.E. Battles over resources and territory among the regional groups resulted in conflicts and, eventually, a stratification of the society and an intertribal slave trade.

Portuguese explorers first reached the Congo River at its Atlantic port in 1482. With slaves in use across the region, the colonists bought their own Congolese slaves to increase their workforce. By 1720, the American colonies, Denmark, England, France, Italy, and other countries were capturing and exporting a combined 15,000 slaves annually from central Africa to the West, including the Caribbean, eastern North America, and South America.

Seeking wealth from natural resources, Belgians settled in the Congo in 1885 and formed a private corporation run by King Leopold II as the only shareholder. They negotiated exclusive treaties with local leaders to mine and log the area. By 1908, the corporation was dissolved; the Belgian government ran the country until its independence in 1960.

Throughout the nineteenth and twentieth centuries, civil and ethnic conflicts have defined the Congo region. In 1964, the Congolese people established the independent Democratic Republic of the Congo (DRC). Further conflict among tribal factions in 1971 led to renaming the country Zaire, which remained under authoritarian rule until 1997, when the former name was reinstated. Two recent wars (1996–1997 and 1998–2003) developed due to ethnic conflict; these wars involved eight surrounding nations and resulted in an estimated 5.4 million deaths. A treaty signed in 2002 between the DRC, Rwanda, and Uganda brought a much-needed reprieve.

In 2007, 62.6 million citizens lived in the DRC, mostly in rural enclaves. Approximately 65 percent of the population was employed in farming, 25 percent worked in an aspect of the mining industry, and 10 percent worked in logging or the forest products industry.

The DRC has substantial mining deposits, including copper, diamonds, gold, minerals, and silver. Overseas interests control the majority of the large mines, paying wages to 350,000 people. Profits are typically exported in the form of refined products.

Against the backdrop of widespread poverty, international mining and logging companies often play the duplicitous roles of community provider and profit-maker. In remote areas, these companies control more than just wages. They build housing, construct schools, and pay for health care facilities. Despite these improvements, insufficient investment stays in the DRC, contributing to ethnic strife.

With per capita annual incomes very low, averaging $311, disputes between employees and management at the large companies are common. For example, in 2008, 500 workers at the Canadian mining company Potash Corporation went on strike in an effort to increase wages and profit sharing. The world's largest producer of fertilizers eventually agreed to some of the wage increases in 2009 in exchange for an end to the strike.

This Mbuti family is shown in their forest camp. At the beginning of each dry season, which normally lasts from December through February, these pygmy people leave their villages and set up camp in the forest, using sticks, vines, and large leaves to build seasonal shelters. *(Randy Olson/National Geographic/Getty Images)*

The different indigenous groups in the DRC speak at least 246 languages and number about 130,000. The Mbuti people of the eastern highlands live in small settlements and are subsistence hunter-gatherers. As pygmies, their compact, strong bodies average 4 feet 11 inches (1.5 meters) tall. They eat freshwater crabs, forest yams, roots, fruits, and nuts. They catch larger game periodically, including the giant forest hog (*Hylochoerus meinertzhageni*), the black wildebeest (*Connochaetes gnou*), and several monkey species.

The Mbuti habitat has been negatively impacted by logging, mining, and conflict among neighboring groups. Two major changes faced by the Mbuti include the loss of large game due to overhunting and the lack of acknowledgment by the government of their ownership of the land on which they have lived for centuries.

Pollution and Damage

Periodic wars in the DRC over the nineteenth and twentieth centuries resulted in widespread damage to both the forests and wildlife. When a peace agreement was reached in 2002 in response to the latest conflict, the country had almost no economy. The government was $4.9 billion in debt and had little income. Per capita annual income had declined over the previous decade.

To increase economic development, the currency—the African franc—was devalued. In order to spur economic growth, commercial rain forest logging increased. Between 1990 and 2006, 4.9 percent, or up to 3,088 square miles (7,998 square kilometers), per year was logged from the DRC. Smaller-scale deforestation included firewood gathering and clearing land for new farming plots.

The Congolese government struggled to regulate the larger forestry industry, and corruption was an impediment. Rather than use consistent and open bidding for tree removals on public land, closed-door meetings often resulted in secret agreements. As soon as a new logging road is built, a spectrum of hunters, squatters, and farmers use the infrastructure to push deeper into the Congolese forests.

Because the war-torn DRC lacks suitable food production industries, a considerable pressure on the rain forests is food availability for the people. The domestic animal population, which included chickens, cows, goats, oxen, pigs, and sheep, was decimated during the fighting. In order to satisfy food demand, a sharp increase in "bush meat" sales occurred, including antelopes, bongos, crocodiles, monkeys, rats, snakes, and wild hogs.

Since 1999, bush meat consumption has soared to 8 pounds (3.5 kilograms) per person per year, far above the historic level of 1 pound (0.5 kilograms). In addition, an international market for bush meat has developed in places as far away as New York and Paris. The consequent increase in hunting has devastated

CONGO: FIVE U.N. WORLD HERITAGE SITES IN DANGER

Site Name	Location	Size (Square Miles)	Features
Garamba	Eastern Woodlands	1,899	Densely forested rolling hills and swamps are home to the only four remaining wild northern white rhinoceroses in the world.
Kahuzi-Biega	Eastern Highlands	2,316	These mostly high-altitude forests support up to 130 mountain gorillas, which live in an area ranging from 4,000 to 5,000 feet (1,219 to 1,524 meters) above sea level.
Okapi	Northeastern Woodlands	5,299	Bordering Sudan and Uganda, this wildlife reserve supports up to 6,300 okapis, a zebra-like mammal that lives in the dense rain forest.
Salonga	Central Basin	13,899	Spanning the central Congo rain forest, this is the largest of the Congo's parks. It supports crocodiles, elephants, monkeys, and other species.
Virunga	Eastern Highlands	3,050	The area contains eight large dormant volcanoes reaching to 15,000 feet (4,572 meters) and includes savannas, swamps, and alpine forests.

Source: List of World Heritage Sites in Danger, United Nations Educational, Scientific, and Cultural Organization, 2008.

wild animal populations. Tropical forest species may be prolific, but most native animals have relatively slow reproduction cycles. The forest elephant, for instance, does not reproduce until about fifteen years of age.

Although banned in 1989 by the United Nations treaty the Convention on International Trade in Endangered Species of Wild Fauna and Flora, poaching in the DRC continues. Elephants and rhinos are killed for their ivory tusks, a practice that has been driven by a new demand for ivory in China. While enforcement of the treaty by United Nations peacekeepers has pushed some sellers to flee to Angola and Sudan, the practice continues.

A single elephant grows two tusks that weigh about 33 pounds (15 kilograms) each. In 2002, ivory sales reached 18 tons (the ivory from more than

545 elephants). Although one tusk may sell for up to $4,000, all a hunter might receive in payment is some clothing, food supplies, and a carton of cigarettes.

Mitigation and Management

The Democratic Republic of the Congo is Africa's second-largest nation. It is bigger than France and Germany combined. Although the rain forest habitat supports hundreds of unique species, this biome has suffered heavily from wars and civil strife.

As of 2009, only 9 percent of the total forests was protected, resulting in illegal hunting, occupation by displaced communities, and unlawful logging. Wildlife—chimpanzees, elephants, giraffes, gorillas, monkeys, and rhinos—often has been harassed and pushed from its home territory, and many animals have been killed.

After the initial implementation of the 2002 peace plan, an international coalition was formed to address these issues. The Congo Basin Forest Partnership, made up of local and international government agencies and private international interests, built a network of resource managers and researchers to monitor and conserve up to 40 percent of the rain forest. These measures increased the links among thirty international conservation organizations, government agencies, and international logging and mining interests. Goals include meeting regularly to share information, using consensus-building models to better manage areas, improving enforcement efforts for protecting habitat and species, and increasing funding for local and state government management operations.

From 1979 to 1996, the Congolese government created several national parks and wildlife reserves to protect rain forest resources, including Garamba (in the eastern woodlands), Kahuzi-Biega (eastern highlands), Okapi (northeastern woodlands), Salonga (central basin), and Virunga (eastern highlands). Although these parks and reserves signaled a framework for better managing the Congo forests, little resources were invested in regulatory and enforcement efforts.

In 2004, the United Nations responded to severe environmental threats to the DRC by placing all five locations (previously designated as World Heritage Sites) on their Danger List. By taking this action, the United Nations committed its resources to enhancing conservation efforts, including the payment of $4.1 million in annual salaries, and alerting the international public to the scope of the problem.

Garamba National Park, for example, contains eastern savanna that is an ideal habitat for the northern white rhinoceros (*Ceratotherium simum cottoni*). This species was estimated at 500 individuals in the 1970s; Sudanese poachers reduced the herd to only thirty-two by 2002. By 2009, only four northern white rhinoceroses could be located. Now classified as critically endangered, extinction is all but certain for this species.

Civil unrest, widespread poverty, and a birthrate above 3.2 percent annually continue to be major hurdles to strengthening conservation efforts. Despite hostilities officially ending in 2002, up to 45,000 people have perished each month due to the consequences of war. Devastation to the country's infra-structure includes a loss of hospitals and schools, poorly maintained roads, little economic activity, and insufficient food and water.

In this postwar environment, less than 1 percent of the deaths is a direct result of violence. Rather, the high mortality rate (81.2 deaths per 1,000 births) in the DRC, which is 57 percent higher than surrounding sub-Saharan African countries, is primarily the result of malnutrition and diseases such as dysentery, malaria, and measles; children are particularly susceptible. The Congolese government spends less than $15 per person per year on health care.

The DRC faces considerable challenges to stabilize its economy and address generations of conflict and unrest. Contributing to the complexity, only 2.5

MEMBERS OF THE CONGO BASIN FOREST PARTNERSHIP

Countries: Belgium, Cameroon, Canada, Central African Republic, Democratic Republic of the Congo, Equatorial New Guinea, European Commission, France, Gabon, Germany, Japan, South Africa, the United Kingdom, and the United States.

International Agencies: Commission in Charge of Forests in Central Africa, International Tropical Timber Organization, The World Bank, and the United Nations.

Private or Nonprofit Groups: African Wildlife Federation, American Forest and Paper Association, Conservation International, Jane Goodall Institute, Society of American Foresters, World Conservation Union, and World Wildlife Fund.

percent of the population is older than 65 years of age (life expectancy is only 54 years), and a high birthrate has led to the average family having six children (the ninth highest rate worldwide in 2009).

Plans to stabilize the DRC hae centered around a countrywide effort to educate a population with a median age of sixteen years. Better schools are emphasized as a way to achieve true peace and economic stability, which, in turn, can help conservation strategies to succeed and better protect the rain forests.

Selected Web Sites

Congo Basin Forest Partnership: http://www.cbfp.org.
Congo World Heritage Sites: http://whc.unesco.org/en/statesparties/cd.
Democratic Republic of the Congo: http://www.un.int/drcongo/.
Central Congo Basin satellite view: http://www.nationalgeographic.com/wildworld/ profiles/photos/at/at0110a.html.
Wildlife Conservation Society-Congo: http://www.wcs-congo.org.

Further Reading

Barter, James. *The Rivers of the World—The Congo.* Chicago: Lucent, 2003.
Blanc, J.J., et al. *African Elephant Status Report.* Gland, Switzerland: The World Conservation Union, 2003.
Giles-Vernick, Tamara. *Cutting the Vines of the Past.* Charlottesville: University Press of Virginia, 2002.
Heale, Jay. *Democratic Republic of the Congo.* Tarrytown, NY: Marshall Cavendish, 1999.
Muruthi, Philip. *African Heartlands.* Washington, DC: Island, 2004.
Perez, M. Ruiz. *Logging in the Congo Basin.* Amsterdam, The Netherlands: Elsevier, 2005.
Prunier, Gerard. *Africa's World War.* New York: Oxford University Press, 2008.
Tayler, Jeffery. *Facing the Congo.* St. Paul, MN: Ruminator, 2000.

8 | Black Forest Germany

Located in a corner of southwestern Germany near its borders with France and Switzerland, the Black Forest contains 2,320 square miles (6,009 square kilometers) of public and private lands. At 105 miles (169 kilometers) long north to south and 25 miles (40 kilometers) wide west to east, this temperate, deciduous forest ranges from an approximate elevation of 2,000 feet (610 meters) in the north and climbs to mountain ridges reaching 4,897 feet (1,493 meters) in the south.

The forest straddles a continental divide. On its western border is the Rhine River, which flows to the Atlantic Ocean. On its eastern side are high plateaus that contain the headwaters of the Danube and Neckar rivers, which flow to the Black Sea.

In German, *Schwarzwald* means "black forest." The name comes from the density of trees that grow in a tight canopy, blocking out most overhead light. This mountainous climate experiences an average winter temperature of 25 degrees Fahrenheit (-4 degrees Celsius) and an average summer temperature of 68 degrees Fahrenheit (20 degrees Celsius) and receives approximately 78 inches (198 centimeters) of precipitation annually. In winter, it may snow up to 100 days, producing 65 inches (165 centimeters). The climate is influenced by a mixture of moist western winds from the Mediterranean Sea and locally cooled mountain air.

BLACK FOREST TREES

Tree Name	Percentage of Forest
Norway Spruce	28
Scots Pine	23
Beech	16
Alder, Birch, and Poplar	10
Oak	9
Ash, Maple	7
Larch	3
Douglas Fir	2
Silver Fir	2

Source: State of Baden-Wurttemberg government survey data.

About 90 percent of the Black Forest is covered by trees; the rest is farmland. The forest hosts 1,980 plant species, of which 620 consist of trees. Prevalent tree species include the Scots pine (*Pinus sylvestris*), common beech (*Fagus sylvatica*), Norway spruce (*Picea abies*), elm (*Ulmus glabra*), and silver fir (*Abies alba*).

Shrubs, bushes, and grasses have adapted to moist and low-sun conditions along the forest floor. Plant species, such as old man's beard (*Clematis vitalba*), climb onto bushes or small trees, seeking better light. A range of insects feed on the clematis's white flowers. The hawthorn (*Crataegus laevigata*) is a large shrub that grows up to 20 feet (6 meters) tall in order to gain suitable sunlight; it produces a small, dark-red, nutrient- and oxidant-rich berry eaten by birds in the winter.

A number of animals also have adapted to thrive in the local mountain conditions. Rich soils support the giant earthworm (*Lumbricus badensis*), which may grow to 2 feet (0.6 meters) in length, feeding on organic matter dropped from trees and plants. The badger (*Meles meles*), known as a *dachs* in German, is related to ferrets and weasels. Named for its distinctive white-colored stripe of fur running from forehead to tail, these short-legged and stocky mammals are fierce predators that live on a diet of voles, ground birds, snakes, rabbits, and squirrels. Badgers have dexterous paws with sharp claws that make them equally capable at catching prey and defending themselves. Their jaws are able to lock closed with great pressure.

Roe deer (*Capreolus capreolus*) are small herbivores that blend well into the forest shades of red, grey, and brown. They browse through dense woods, eating small plants, young trees, and bark. Active in the early morning and evening, roe deer also frequent areas where thick woods meet open grasslands.

The Black Forest sits on one of Germany's oldest geologic formations. Approximately 250 million years ago, its northern reaches consisted of a shallow basin that flooded regularly. Layers of sedimentary sandstone and limestone rock up to 300 feet (91 meters) thick formed in this region. In its southern region, a collision of tectonic plates 65 million years ago pushed the granite and gneiss bedrock upward to form mountain peaks cresting with the Feldberg Mountain at 4,898 feet (1,493 meters). Years of glacial erosion and weathering from wind and rain have worn the peaks down by several hundred feet.

Human Uses

The Black Forest region was settled in 2500 B.C.E. by tribes that emigrated from Scandinavia. They found a forest consisting of 95 percent beech trees, rich soil in wide river valleys, and moderate winters. Later, in 9 C.E., the Romans tried to invade this region but were pushed back by a Germanic tribe called the Cherusci. They and other tribes, including the Bructeri, Chatti, Chauci, Marsi, and Sicambri people, raised domesticated cattle, maintained small farm plots, and logged the forest for building materials and firewood.

After the Roman Empire collapsed in 476 C.E., the Germanic tribes continued to expand, eventually unifying into a large enough army to reclaim lands from the Romans by 650 C.E. In 800 C.E., Charlemagne (Charles the Great), King of the Franks (a collection of tribes from eastern and central Europe), established the Carolingian Empire, which included the Black Forest.

During the thirteenth century, regional economic growth led peasant settlements to clear 70 percent of the forested lands to supply charcoal and wood to growing European cities. In the 1500s, mountain hot springs were channeled and their waters were used as a healing treatment, which drew many visitors to the mountains.

During the Industrial Revolution in the eighteenth century, logging activity peaked and was mostly unregulated; piles of trees were dragged to rivers and rafted downstream. Wood from the Black Forest was used to build entire communities, to construct bridges, to provide structures for mines and tunnels, and to make other wood products that were distributed across Europe.

In the nineteenth century, the local mining industry developed. Copper, iron, and silver deposits provided new wealth and economic growth to the region. The nation of Germany was established in 1871.

After several regional conflicts, two world wars, and the lengthy cold war, today's Germany is a global industrial power with a prominent international role in the United Nations and a regional one in the European Union. Germany

(© Goos_Lar/Fotolia)

RETURN OF THE LYNX

The lynx *(Lynx lynx)* is a medium-sized cat species that was once abundant in the Black Forest. A skilled hunter with excellent hearing, smell, and sight, this feline can grow to a weight of up to 65 pounds (29 kilograms) and consumes small to medium animals, including mice, hares, and deer. By the 1880s, the lynx had been overhunted for its valuable fur coat, and it disappeared from the region.

In 1991, a group called the Lynx Initiative proposed reintroducing these cats to restore the historic ecological balance to the biome. Hunters and farmers, who felt that domestic animals such as chickens and sheep, as well as game animals such as roe deer and boar, would be attacked if the lynx were reintroduced, protested vigorously to the state government, placing the project in a bureaucratic stalemate.

While conservation and agricultural groups were debating the merits of such a proposal, the lynx moved into the Black Forest on its own, arriving from eastern France and northern Switzerland. Although no formal count exists, conservationists believe that approximately a dozen individuals roam the darkened woods along the trails that their ancestors walked two centuries ago.

houses Europe's largest population with 82.4 million residents; nearly 86 percent resides in large cities.

The Black Forest sits within the state of Baden-Wurttemberg, which has some 10.7 million residents. Inside the forest boundaries, there are three-quarters of a million people. Its rural character is largely intact. Homes that surpass 200 years in age are common. The local economy still relies on the forest, and selective logging is performed, both for building materials and firewood.

Small farms grow produce and raise livestock. Local businesses include a spectrum of health spas and tourism endeavors, and independent crafts workers manufacture goods such as cuckoo clocks, watches, and glass items.

Pollution and Damage

Although dozens of generations of Germans have cut trees, mined for resources, and grazed livestock in the Black Forest region, two human activities in particular applied the greatest pressure on the ecosystem. The first is ill-conceived forest management practices that stretch back to the thirteenth century. The second relates to past and present industrial pollution that has resulted in an accumulation of toxins in animal and plant tissue as well as in soils.

Black Forest landowners historically have harvested their forests by selectively removing individual trees, a method called *plenterwald* in German. This practice limited ecological impact and ensured a steady income.

When European demand and prices for trees peaked in the 1820s, plenterwald was viewed as not economical and inefficient, and a new method, called *baden,* was introduced. Baden required larger swatches of forest to be clear-cut. This new system resulted in sections of leveled forest, leaving bare slopes next to standing forests, an early example of clear-cut.

Although profits soared, mudslides became common during rainstorms and debris poured into the Rhine and other rivers. This scarred landscape made it difficult for new trees to naturally seed and establish themselves, and deer often consumed emergent tree saplings. Water quality, especially in nearby wells, declined due to sediment and turbidity.

In 1870, local farmers began protesting these damaging logging practices, noting the ecological impacts. In response, the government undertook a multi-year effort to replant cutover slopes. Instead of slower-growing, native beech, fir, or oak trees, however, most of the seedlings planted were highly competitive Norway spruce trees (*Picea abies*).

This monoculture was another misstep in forest management. Although the spruce thrived in the temperate climate, their overabundance had immediate drawbacks. The trees were planted too closely together, forming a dense canopy that allowed little light to pass through to the ground. Previously drier and open forests remained damp. Understory grasses and plants shrunk in density. Sunlit paths were transformed into muddy ditches.

The majority of the spruce trees grew shallow roots in the mountain soil. Up to 11.6 billion cubic feet (0.3 billion cubic meters) of trees in the Black Forest were knocked down by windstorms between 1967 and 1999, and up to 13,590 acres (5,500 hectares) were burned by forest fires between 1981 and 1999. Most of this loss was due to overcrowded spruce plantings.

Beginning in the nineteenth century, the industrial revolution brought great economic expansion and prosperity to Germany. It also resulted in considerable emissions from European factories that deposited airborne and water pollution into the forests, causing trees to be stunted, drop foliage, and sometimes die.

One of the most common environmental impacts of industrialism, recorded in 1852, is a phenomenon called acid rain. It occurs when airborne pollution that is high in nitrogen and sulfur mixes into the higher atmosphere and is then precipitated. Common sources of such pollution are coal-fired power or industrial factories.

Acid rain impacts forests by starving them of nutrients. Liquid sulfuric acid saturates tree roots and prevents the trees from absorbing calcium and magnesium. Concentrated metals such as iron, lead, and zinc also reduce the beneficial fungi that live on trees and normally help them absorb minerals and moisture. The end result after years of acid rain is a barren forest with little surviving vegetation.

In 1983, Black Forest managers estimated that 70 percent of the trees had been impacted by acid rain. The symptoms varied with the exposure, weather, rainfall, location, and tree size. Surveys in the 1990s found that up to 80 percent of the silver fir and 50 percent of the spruce trees were damaged. By 2004, Black Forest tree losses were estimated at $80 million annually, or one out of every four trees.

An ecological assessment performed in 2004 by the federal agricultural ministry found widespread "dead zones" in Black Forest terrestrial and aquatic communities. The cause was not only years of acid rain, but also newer industrial emissions containing ammonium, nitrates, nitrogen, and other acids. The source was the tall smokestacks in the Rhine Valley, Germany's economic heart. This region, located approximately 50 miles (80 kilometers) to the west of the Black Forest, has Germany's highest concentration of power plants, factories, and industrial plants.

The towering exhaust stacks send emissions high into the air via weather systems and deposit them in the form of precipitation across the Black Forest. The 2004 sampling found that a blanket of this toxic brew had impacted even the smallest soil-dwelling microbiological bacteria. Without these bacteria colonies, species such as the ground beetle (*Carabidae sp.*) have insufficient food. This impacts species farther up the food chain, including insectivore birds such as the black-capped chickadee (*Poecile atricapillus*).

Mitigation and Management

Acid rain has been one of the biggest hurdles for Black Forest managers, because it does not originate in the southwestern region. A primary step toward reducing this phenomenon was Germany's signing of the 1979 Convention on Long-Range Transboundary Air Pollution created by the United Nations.

This treaty, which included fifty-one parties from across the Northern Hemisphere, uses eight protocols to limit industrial air pollution. The core goal of the convention is to share scientific advances in the field of emissions controls, such as installing high-pressure water scrubbers and other equipment to capture pollutants before they are released into the air. The Convention has recorded a decline of annual emissions of sulfur dioxide from 60,000 kilotons in 1980 to about 15,000 in 2008.

Another cleanup effort affecting the Black Forest is planet-wide. In 1997, Germany signed the Kyoto Protocol, an international environmental treaty established to stabilize worldwide greenhouse gas concentrations in the atmosphere. Greenhouse gasses include carbon dioxide, methane, nitrous oxide, ozone, and water vapor. Although naturally present in the atmosphere, man-made pollution has caused a surge in concentration, resulting in trapped heat planet-wide, leading to climate change.

Because much of the air pollution that affects the Black Forest originates south and west of it, participation from countries such as France, Greece, Italy, Portugal, Spain, Switzerland, and the United Kingdom is considered critical to a region-wide solution. Between 1990 and 2004, a handful of nations lowered their greenhouse gas emissions, including France (-0.8 percent), the United Kingdom (-14 percent), and Germany (-17 percent). During the same period, however, several countries increased their pollution, including Greece (+27 percent), Portugal (+41 percent), and Spain (+49 percent). These countries, as signatories of the treaty, will need to meet an 8 percent reduction goal by 2012 as part of the Kyoto Protocol.

Between 1820 and 1970 much of the Black Forest was clear-cut to support German economic development. It was not until 1986 that forest managers formally halted clear-cuts and returned to historic practices of small selective tree removals.

In 2002, 105 local communities in the Black Forest region formed the Schwarzwald Nature Park. This 1,447-square-mile (3,748-square-kilometer) preserve encompasses cities and towns with 715,000 residents. A regional forestry commission oversees any logging inside the preserve. The overarching aim of the park is to maintain biodiversity and support efforts toward a sustainable future for the Black Forest. These efforts include conserving and protecting the beauty and character of the landscape, improving its recreational and tourist value, and fostering traditional agriculture.

There are several state-paid foresters who patrol the Black Forest region and implement conservation-based regulations. By logging under the selective system, the cutting process may take longer, so as not to harm nearby trees, but the end result is a thriving forest biome. Cuts in the winter are limited, due to erosion impacts when vegetation is dormant. To support these practices, Black Forest landowners have banded together and formed a cooperative for selling wood. This group has 3,500 landowners with 17,791 acres (7,200 hectares) of forest.

With fewer large tree clearings, the local economy has shifted to relying on tourism. The region recorded 28 million overnight stays in 138,000 hotel beds and 60,000 private rooms in 2006. Skiing flourished in 1907, with the opening of twenty-eight ski lifts on the Feldberg, and continues to this day. Private and public groups maintain a network of 14,291 miles (22,994 kilometers) of hiking trails across the region.

This temperate deciduous forest in southwestern Germany has been the source of significant economic livelihood over the last 2,000 years. Through notable government environmental regulations and land protection efforts, historic practices of overutilization and pollution in the forest appear to be on the decline, allowing this natural biome to thrive once again.

Selected Web Sites

German Federal Environment Agency: http://www.umweltbundesamt.de/index-e. htm.
Johann Heinrich von Thünen-Institut, German Federal Research Institute for Rural Areas, Forestry and Fisheries: http://www.vti.bund.de/en/.
Schwarzwald Nature Park: http://en.naturparkschwarzwald.de.
Southern Black Forest Nature Park: http://www.naturparksuedschwarzwald.de/en/index_en.php.

Further Reading

Bekker, Henk. *The Black Forest.* Walpole, MA: Hunter, 2007.
Evans, Julian. *The Forests Handbook: Volume 1.* Malden, MA: Blackwell, 2001.
Kuusela, Kullervo. *Forest Resources in Europe.* London, UK: Cambridge University Press, 1995.
McEvoy, Thom. *Positive Impact Forestry.* Washington, DC: Island, 2004.
Philipp, Dorthee. *Schwarzwald: Black Forest.* Berlin, Germany: Art Stock, 2007.

9 Borneo Rain Forest Brunei, Indonesia, and Malaysia

Located in the South Pacific, Borneo sits near Sumatra to the west and the Philippines to the northeast. The island, comprised of the countries Brunei, Indonesia, and Malaysia, has one of the largest contiguous protected tropical forests in the world. The Heart of the Borneo Forest spans 85,000 square miles (220,150 square kilometers).

The majority of the population of Borneo, over 18 million residents, lives in coastal communities, leaving the central forest less developed. This heartland is an impressive biome. It contains many endemic animal and plant species that do not exist anywhere else in the world. The island also contains approximately 10 percent of the world's rain forests.

Part of a chain of 25,000 islands, Borneo has an uncommon geological past. While the last 100 million years of tectonic plate movement have pushed neighboring islands, such as the Philippines and New Guinea, miles across the ocean, Borneo has stayed within an equatorial belt, 1° north of the equator. This has resulted in a unique biotic range of species that have not had to adapt to climate changes.

Borneo's tropical climate features two rainy seasons: a monsoon season with weather blowing in from the northeast between December and March and a second season with winds shifting around from the southwest from May to October. Annual rain levels range between 110 and 196 inches (279 and 498 centimeters), with temperatures averaging around 75 degrees Fahrenheit (24 degrees Celsius) in the lowlands and 63 degrees Fahrenheit (17 degrees Celsius) in the mountains.

Forest Types

Two hundred years ago, Borneo was almost covered with dense and mostly impassable rain forests. Today, the forested area has been reduced by 50 percent, although the central forest still resembles that of the original biome. The Heart of Borneo contains two types of montane forests, lowland and highland.

The island forests boast remarkable diversity. The central mountains host a bounty of plants and wildlife, including 15,000 species of plants, more than 2,900 tree species, 420 bird species, 394 fish species, 222 mammal species, and 150 reptile and amphibian species. In just a 24-acre (9.7-hectare) sample, 700 tropical tree species can be found, a greater density and diversity than any other highland forest.

The lower montane forest is dominated by dipterocarp tree species, named from the Greek word for "two-winged," which describes the paired fruit these trees produce. Scientists have identified up to 284 dipterocarp species, some growing 230 feet (70 meters) high. When these trees reach full fruit production, the tropical forest overflows with food for everything from elephants to orangutans.

In the slightly higher elevations, tropical oak trees (*Quercus subsericea*) are common. This hardwood species produces acorns that are eaten by area birds and mammals. Housed in a durable shell, the acorn heart is packed with meat rich in carbohydrates, as well as tannins, which aid in the absorption of minerals and nutrients. Some bird, deer, and mouse species in Borneo rely on acorns for 35 percent of their diet.

Wildlife

Although Borneo is an island—the world's third largest after Greenland and New Guinea—its population of large animals resembles that of a much larger landmass.

Borneo pygmy elephants (*Elephas maximus borneensis*) are about 6 feet tall (1.8 meters), 3 feet (0.9 meter) shorter than Asian elephants, and roam the northeastern lowlands. They can eat up to 330 pounds (150 kilograms) per day of grasses, palms, and fruits such as the wild banana. Thought to have lived

The Proboscis Monkey, or long-nosed monkey (*Nasalis larvatus*), is native to Borneo. The nose is thought to help attract females, but it also assists in emergencies. When the monkey becomes agitated, the nose swells in size, which amplifies its warning calls. (© emprise/ Fotolia)

on the island for 300,000 years, this species has evolved to be more docile than their Asian and African counterparts.

The endangered Sumatran rhinoceros (*Dicerorhinus sumatrensis*) is the smallest of the five global rhinoceros species, with an adult weighing on average 1,600 pounds (726 kilograms) and standing about 5 feet (1.5 meters) tall. Using a system of trails to move through the thick forest, this mostly solitary species grazes on fruits, leaves, and saplings, as well as aquatic plants. In order to supplement this diet, the rhinoceros frequents mineral- and salt-laden springs.

The Bornean orangutan (*Pongo pygmaeus*) is a highly intelligent large primate native to the island. Humans are closely related to the orangutan, with up to 98 percent similar DNA. Adult orangutans are the biggest arboreal animal in the world, reaching 5 feet (1.5 meters) tall and weighing up to 180 pounds (82 kilograms). Early Indonesians felt that orangutans were so like themselves that they named them "orang hutan" meaning "forest people." This endangered species lives in trees at elevations of up to 4,500 feet (1,372 meters), relying on the trees' upper levels for food, shelter, and travel lanes. The animal's diet consists of a diversity of vegetation, including leaves, bark, and flowers, and more than 200 kinds of forest fruit.

In 2006, World Wildlife Fund scientific researchers identified up to fifty-two new species of animals and plants in Borneo, including thirty new fish species, two types of frogs, and several types of plants. A three-tenths-of-an-inch-long (7.62-millimeter-long) fish, called *Paedocyopris micromegenthes*, was one of the findings. Not bigger than the tip of a fingernail, scientists believe this could be one of the world's smallest living vertebrates. Researchers also discovered a catfish, *Glypothoras exodon*, which uses small suction-like appendages to hold on to rocks in swift river currents.

Preying mantises (*Mantodea sp.*) have two grasping, spiked forelegs in which prey is caught and held securely. There are thousands of species of these predatory insects. Smaller species eat other insects; larger ones prey on small and often young lizards, frogs, snakes, birds, and even rodents. Preying mantises also may engage in cannibalism. *(Tim Laman/National Geographic/Getty Images)*

Human Uses

A little larger than the state of Texas, the island of Borneo has almost 18 million residents, most of whom live in Indonesia and Malaysia. Due to significant economic expansion in the last two decades of the twentieth century, the island-wide population has grown from 9 million in 1980 to over 18 million in 2009.

Borneo's economy relies on its natural resources. Up to 60 percent of island residents is employed in the agricultural sector, 15 percent of which works growing and processing the seeds of oil palm plants. Roughly 2.5 acres (1 hectare) of palm plant produces enough seeds to provide 1,585 gallons (6,000 liters) of palm oil in a single crop.

Logging is a significant segment of the Borneo economy, and it is the third largest export industry after oil and gas, generating $9 billion in earnings in 2002. Trees such as the dipterocarp species are cut to make knot-free wood for

BORNEO POPULATION

Country	Percentage of Total Borneo Population	Number of People
Indonesia	67	12.5 million
Malaysia	31	5.8 million
Brunei	2	0.3 million

Source: Government Web sites of individual countries and provinces.

buildings, furniture, and products such as plywood, resins, varnishes, and veneers. International logging companies from China, Japan, and the United States have a presence in Borneo, demonstrating the strong demand for tropical woods.

A number of products are harvested in the forest for sale worldwide. These include the various tree seeds and beans that are picked and used as ingredients in manufacturing cosmetics and chocolate. Another example is the rattan palm. The branches of this tree are cut, dried, and used to weave fabric and other goods. The resulting products, which range from handbags to rugs, are both lightweight and durable.

Borneo is occupied by more than thirty groups of natives known collectively as the Iban. Numbering around 2.2 million, they have lived for thousands of years on the island. When European explorers first encountered these tribes, they referred to the Iban people as "Sea Dayaks," because they operated hostile pirate boats along the northern coast. The Iban made a particular impression on the Europeans: They used head-hunting as part of their practices of war against encroaching groups.

While these peoples have adapted to modern society, they maintain many of their ancestral traditions, including living in dwellings called longhouses, which host up to 200 people for events. These communal structures demonstrate the communal aspect of Iban society: Individuals work together to harvest rice, catch fish, pick fruits, and grow vegetables. Most Iban settlements are found along river corridors, where there is access to freshwater for daily needs and travel.

Pollution and Damage

Much of the forest damage on Borneo occurred during the last decade of the twentieth century and has continued through the first decade of the twenty-first century. This damage has primarily happened on the outer two-thirds of the island, close to growing cities and towns.

For the most part, the heart of the island has been left relatively undisturbed due to geography and limited infrastructure investment. The central region highlands are not readily accessible, and limited government spending has meant few developed roads. The heartland also is situated between Indonesia and Malaysia, farthest from population and commerce centers.

Both Indonesia and Malaysia have, however, rapidly increased their lowland logging since 1985. By 2007, 68 percent of the lowland rain forests in Indonesia and 56 percent in Malaysia were heavily cultivated. And this widespread deforestation continues to move inland toward these central forests.

When the Indonesian government moved to limit widespread logging in 2001, they imposed size restrictions on land parcels. These regulations forced private loggers into smaller districts, but this effort failed as no limits were placed on the number of parcels a company could harvest. Instead of reducing deforestation, the regulation had the opposite effect, pushing companies farther into virgin forest.

Since 2005, the rate of lowland deforestation has increased to 5,019 square miles (12,999 square kilometers) per year. Some logging companies skirt the laws regulating logging. One illegal practice involves cutting logs in Indonesia and then hauling them to Malaysia and relabeling them to indicate they were harvested using sustainable practices for sale in the U.S. or European markets.

Logging companies, as a measure to ensure the sustainability of the rain forest, have just begun the practice of replanting cutover areas with seedlings. Large clear-cuts have left the landscape full of debris that dries in the powerful tropical sun. A 1997 drought turned a large clear-cut area in Indonesia into a tinderbox, and a resulting forest fire consumed 25 million acres (10 million hectares).

Aggressive clear-cutting poses threats to the survival of many rain forest species. The large animals, often called mega-vertebrates because of their size and the attention drawn to them by the public, are in particular danger.

For example, the forest herd of Sumatran rhinoceros in the northeast corner of Borneo has shrunk to fewer than fifty animals. Hunters periodically have shot these animals for their horns, thought to have medicinal value, though this is scientifically unproven. In 2007, the Malaysian government used a hidden camera to record the first known rhinoceros in the central forest region. This discovery has resulted in intensified efforts to better protect this central forest.

In 2008, the number of pygmy elephants in Borneo was estimated at less than 1,200; this count was down from around 3,000 in the early 1990s. The pygmy elephant has lost critical habitat to logging, forcing this species to live in reduced territories under considerable stress.

DECLARATION ON THE HEART OF BORNEO INITIATIVE
Three Countries, One Conservation Vision

We, the Governments of Brunei Darussalam, Indonesia and Malaysia, recognizing the importance of the Island of Borneo as a life support system, hereby declare that:

1. With one conservation vision and with a view to promote people's welfare, we will cooperate in ensuring the effective management of forest resources and conservation of a network of protected areas, productive forests and other sustainable land-uses within an area which the three respective countries will designate as the "Heart of Borneo (HoB)", thereby maintaining Bornean natural heritage for the benefit of present and future generations, with full respect to each country's sovereignty and territorial boundaries, and also without prejudice to the ongoing negotiations on land boundary demarcation.

2. The HoB Initiative is a voluntary trans-boundary cooperation of the three countries combining the stakeholders' interests, based on local wisdom, acknowledgement of and respect for laws, regulations and policies in the respective countries and taking into consideration relevant multilateral environmental agreements, as well as existing regional and bilateral agreements/arrangements.

3. We are willing to cooperate based on sustainable development principles through research and development, sustainable use, protection, education and training, fundraising, as well as other activities that are relevant to trans-boundary management, conservation and development within the areas of the HoB.

To support this Declaration, we, the three countries will prepare our respective project documents incorporating the strategic and operational plans, which will form the basis for the development of our road map towards realizing the vision of the HoB Initiative.

Source: World Wildlife Fund, 2008.

The orangutan population also has shrunk; it was estimated at less than 48,000 in 2009. These primates rely on adopting the complex social habits of their families. Young primates spend up to their first eight years with their mother, learning how to survive far above the ground in a rain forest. The slow pace of reproduction for this highly intelligent species makes it difficult for them to maintain their numbers while losing habitat.

With increasing demand for energy alternatives to fossil fuels, Borneo has been expanding its plantations of oil palm (*Elaeis oleifera*) plants in order to generate engine fuel and other consumer products. The land that has been cultivated for oil palms has expanded rapidly, growing seventeenfold between 1985 (2,316 square miles, or 5,998 square kilometers) and 2007 (38,610 square miles, or 100,000 square kilometers).

In 2007, Borneo grew 47 percent of the world's supply of oil palm, with land devoted to this production spread across 7,722 square miles (20,000 square kilometers) of farms. After a lumber company clears a section of forest, it has become common to establish a palm plantation; sometimes this is done illegally on previously designated conservation land.

Mitigation and Management

Rapid population and economic growth since the late 1980s has resulted in considerable deforestation. These changes directly threaten the remaining rain forests in the central mountains. In 2001, The World Bank reported that, at current deforestation rates, 85 percent of the lowland forests would be cutover by 2010, and that almost no forests of considerable size would remain outside protected areas by 2020. As early as 2010, the cutover lowland forests verge on 50 percent.

The challenge of protecting and managing forests on an island-wide scale is new to Bornean government leaders. In 1992, the governments of Brunei, Indonesia, and Malaysia commenced high-level discussions at the United Nations Conference on Environment and Development in Rio de Janeiro, Brazil, to address the impacts of deforestation. By 2005, these nations had established planning groups, made up of dozens of intergovernmental agencies, to develop a vision for an area designated as the Heart of Borneo.

In 2006, forest conservation was a central issue at a meeting of the Association of Southeast Asian Nations, a group of regional governments that cooperate to address economic, social, and environmental issues. In 2007, the nations of Brunei, Indonesia, and Malaysia agreed in general terms to protect an 85,000-square-mile (220,150-square-kilometer) parcel in the central mountains (roughly the size of the United Kingdom). This agreement, which protects one-third of the island, provides a blueprint for collective planning and conservation efforts. One key element of this agreement is the government crackdown on the illegal cutting and transportation of logs between Indonesia and Malaysia.

Despite some success in improving conservation and forest management practices, the pace of logging has not shown significant signs of slowing. China, with the strongest worldwide economic growth in the last three decades and a burgeoning middle class, developed a $30 billion logging investment with the Indonesian government in 2006. The investment includes funds to construct hundreds of miles of roads to better transport resources such as timber and palm oil. This injection of Chinese money into a poverty-ridden nation resulted in forest removals to provide plywood, furniture, and other products for a growing mainland population.

Recent controversy erupted in 2007 in the international community when a $1 billion rush order for 27 million cubic feet (1 million cubic yards) of brown tropical hardwood was placed by China to be used for the construction of sports facilities for the 2008 Olympics in Beijing, resulting in up to 6,949 square miles (17,998 square kilometers) of deforestation. In a separate action late in 2007, the United Nations Convention on International Trade in Endangered Species of Wild Fauna and Flora raised the issue of the rapid depletion of Bornean merbau trees as a threat to multiple endangered rain forest species.

Selected Web Sites

Borneo's Moment of Truth: http://ngm.nationalgeographic.com/2008/11/borneo/white-text.html.
Government of Sabah, Forestry Department: http://www.forest.sabah.gov.my/.
Orangutan Organization: http://orangutanfoundation.wildlifedirect.org.
Trees and Shrubs of Borneo: http://www.phylodiversity.net/borneo/delta/.
World Wildlife Fund, Heart of Borneo Forests: http://www.panda.org/about_wwf/where_we_work/asia_pacific/our_solutions/borneo_forests/index.cfm.

Further Reading

Butler, Rhett. *Borneo.* San Francisco: Mongabay, 2007.
Garbutt, Nick. *Wild Borneo.* Cambridge, MA: MIT Press, 2006.
Moeliono, Moe, ed. *The Decentralization of Forest Governance.* London, UK: Earth-scan, 2008.

Russon, Anne. *Orangutans: Wizards of the Rain Forest*. Buffalo, NY: Firefly, 2004.

Schilthuizen, Menno. *Borneo's Botanical Secret*. Jakarta, Indonesia: World Wildlife Fund, 2006.

Wadley, Reed, ed. *Histories of the Borneo Environment*. Leiden, The Netherlands: Kitlv, 2006.

White, Mel. "Borneo's Moment of Truth," *National Geographic Magazine*, November 2008.

World Conservation Union. *Gunung Mulu National Park: Borneo, Malaysia*. Gland, Switzerland: World Conservation Union, 2000.

FORESTS
CONCLUSION

10 | Forest Challenges

The last two centuries have revealed a range of complex environmental challenges to forest communities across the planet. The sources of the problems have been factors such as varying weather systems, climate change, water diversions, water pollution, acid rain, desertification, logging, road or infrastructure construction, farm expansion, and even warfare. Combined, these impacts are exacting a considerable toll on forests planetwide.

Human population growth and widespread industrial development have brought an unprecedented thirst for forest products, which provide the raw materials for a range of everyday items, from flooring to food to fuel for combustion engines and power plants. The rate of forest resource consumption seen in the twentieth and early twenty-first centuries, although managed in some areas with government regulation, is not sustainable in the long term. Up to 1.6 billion people rely on local forests for income, fuel, and food.

Around the world, forests today are vastly smaller, up to 50 percent, than those of preindustrial times. The rates of deforestation have varied over time, but in 2000 the World Resources Institute estimated that logging, farming, and land development were reducing forested areas at a pace of 86,800 square miles (224,812 square kilometers) annually, out of a total of 13 million square miles (34 million square kilometers) of forests worldwide.

At this rate, more than 60 percent of all forests existing today will be gone in less than a century. Despite these trends, there are unique ways in which communities are trying to better manage, conserve, and protect their forests for the future.

Illegal Logging

Addressing illegal logging, especially on land that is publicly owned, is a constant challenge. Limited resources are a perennial issue for governments, even those that have established forestry agencies. Common hurdles include insufficient staff and resources, weak policies and laws, and poor implementaion of enforcement goals.

An example of a nation struggling with such issues is Madagascar. In 1950, tropical forests covered 25 percent of the island nation, which is located 155 miles (249 kilometers) off the coast of East Africa. Up to 90 percent of the island's tropical forest has been leveled since that time.

Although remote, this island's forests contain up to 5 percent of the world's species, including up to 8,000 flowering plants. Decades of logging to convert forests to farms across the country's 226,500 square miles (586,635 square kilometers) have brought widespread negative impacts. Per capita income on the island is very low (around $740 per year), and incentives to raise cash earning crops such as coffee and livestock continue to spur additional, extensive logging.

Between 1945 and 2003, forest removal rates in Madagascar averaged 1.5 percent of the total forest per year. Aggressive cutting reduced the forested areas to just 15 percent by 1999 and then 10 percent by 2009. Although the forest management agency in Madagascar oversees some 39,768 square miles (102,999 square kilometers), it has had insufficient resources and too few employees for decades, making enforcement of illegal logging and poaching of endangered species very difficult.

As a result of government initiatives and international partnerships, the island had forty parks and reserves covering 2,702 square miles (6,998 square kilometers) in 2009. However, despite concerted efforts by the government to control deforestation, up to 772 square miles (1,999 square kilometers) of tropical forest continue to be cut illegally each year. Few resources are available to inventory, let alone stem, such illegal practices.

Seeing the Forest from Space

Before the 1970s, there was little technology available to efficiently and accurately inventory global forests. Any survey required aerial photographic flights and follow-up field inspections. Results would take months to complete.

When the United States launched its Landsat 1 satellite in July 1972, images from space amazed the public. Sent into a near polar orbit, this aircraft could take stunningly detailed pictures and radiometric scans and, by circling in the

Log driving, the process of floating logs down a river to a sawmill, was replaced as the primary mode of transporting lumber with the invention of the railroad and the building of roadways for trucks. However, this practice still is common in parts of Canada. (© Feng Yu/Fotolia)

right orbit, stay in sunlit space almost twenty-four hours per day. By 1985, a network of sophisticated, low-orbit satellites were circling Earth. The first comprehensive global forest surveys tracked the health and condition of forests across the planet.

One example of a collaborative effort to better track forest removals included a partnership between the U.S. National Aeronautics and Space Administration (NASA) and the University of Maryland in an initiative called the Humid

Tropical Deforestation Project. Begun in 1998, this effort uses high-resolution images to track widespread tree removals in three areas over time. Images from several time periods between the 1970s and the 1990s were compared to gauge changes in forest communities in the Amazon Basin, Southeast Asia, and central Africa. These maps have been made available to local governments.

Because it would cost millions of dollars for a single country to fund the design, construction, launch, and maintenance of its own satellites, NASA encourages governments around the world to access and use its images. The most recent satellite, Landsat 7, was launched in April 1999, and this seventh-generation craft provides advanced cloud-free images and three-dimensional imaging. In 2011, the Landsat program, run by the U.S. Geological Survey, is planning to launch the next generation Landsat 8 satellite, which will be equipped with more advanced technology to survey the planet.

Countries such as Madagascar have benefited from satellite photographs of their forests. A study performed in 1990 by Washington University researchers in St. Louis, Missouri, tracked the changes in Madagascar forest cover over the course of the previous thirty-five years. The data showed a 90 percent reduction of forests and identified the logging "hot spots." This information helped to direct limited government resources to address illegal logging. New satellite information can be prepared on a regular basis to better direct fieldwork.

Another example of the effectiveness of satellite imaging was the comparison of images of a section of central Mexican tropical forest taken in 2004 and 2008. The images presented evidence of illegal logging that resulted in clear-cuts in a region designated as a state and local forest and a United Nations Biosphere Reserve.

This particular forested region is the winter home to tens of millions of monarch butterflies (*Danaus plexippus*) who spend the colder months clustered in fir and pine trees. These butterflies, which migrate across North America in the summer months, rely on the shelter of the conifer trees to protect them from freezing or dehydrating during the winter months. Given the stark evidence of the satellite images, public concern for this species has been elevated, and local conservation organizations are working with Mexico's local and state governments to improve oversight of the forests.

Surge in Farming

As the worldwide human population continues to grow by up to 82.8 million people each year, so have the number of farms, especially in tropical areas where the human growth rate is approximately 3.5 percent annually. To meet

A significant portion of this farmer's land has been lost to severe river erosion. Heavy rains have resulted in high runoff, depleting the soil of surface plants, woody matter, and valuable nutrients. Scored by gullies, the land has become bare and unproductive. *(Photo by Mike Goldwater/Getty Images)*

economic demands, food productivity is expected to grow by 60 percent by the mid-twenty-first century. This ever-increasing need has resulted in fewer forests, because many logged parcels have been converted to farmland.

Since the mid-1800s, roughly 965,255 square miles (2.5 million square kilometers) of temperate forests and 2.3 million square miles (6 million square kilometers) of tropical forests worldwide have been converted into cropland and pastures for domesticated animals.

This conversion leads to many environmental changes. Once forests are cut down, they often are cleared of woody debris and leveled. Wind and rain then erode valuable topsoil. Crop irrigation often results in high levels of salts and other minerals in the soil, making it unsuitable for new crops or forest regrowth. Intensive farming lowers soil nutrients, such as phosphorus, potash, and nitrogen. Excessive pesticides, herbicides, and fertilizers are aplied and often accumulate in water supplies. Farm irrigation wells lower the water supply in a portion of an aquifer, reducing the total amount of water available.

After a forest is cleared and farmed using unsustainable methods, the landscape becomes irreversibly degraded. In India, up to one-third of farmland has been overused to the point where it can no longer be tilled for food growth. In South America, nearly 50 percent of forested areas cleared for farming becomes useless after ten years and is abandoned.

Across Southeast Asia, monoculture crops (planting a single species such as bananas, rice, or soybeans in a given area of farmland) make up 65 percent of farms; demand for these crops has resulted in fewer local crops. Larger farms, often subsidized by corporate companies, have used seeds that have been genetically modified to grow in various climates, putting pressure on resources such as local water supplies.

Monoculture crops require high levels of herbicides and pesticides and often are planted adjacent to forests. Widespread olive farming across Europe has been criticized for reducing surface water and groundwater for local forests, and the runoff of herbicides and pesticides from olive groves has polluted stands of trees and watersheds.

The rapidly growing world population has applied significant pressure on the productivity and size of farms. Many small farms have been consolidated to form larger operations. Labeled "factory" or "corporate" farms, these ventures often are owned by multinational companies or large business conglomerates. The United States had at least 2.2 million farms in 2008. Of that number, 177,840 large farms, or just 8 percent of the total, exceeded an average size of 767 acres (310 hectares), while some 1.3 million small farms make up 62 percent of the total farms, located on an average footprint of 110 acres (45 hectares). The large farms, however, earned $111 billion in 2000, while the small farms earned only $20 billion.

With the main goal of elevating productivity and profit, most of these businesses apply methods that are cheapest and most efficient in the short term, without concern for long-term results. The emphasis on more crops and livestock per acre, however, has historically entailed using environmentally degrading methods.

Good Agricultural Practices

In order to better protect forests, watersheds, and other natural areas, a number of procedures and regulations called Good Agricultural Practices were developed, beginning in Europe and North America in the late 1990s. Formalized by governments and agricultural industry representatives to improve crop production, increase food safety, limit pollution, and enhance efficiencies, these practices are now in use worldwide.

In 2003, the United Nations Food and Agriculture Organization adopted its own set of Good Agricultural Practices. These are grouped into several categories: animal or crop production, energy and waste management, human health and welfare, soils, water, and wildlife and landscape.

MAINTAINING A BALANCE

One of the environmental assets of a forest is its biodiversity (the measure of living organisms and plants in an ecosystem, often used as an indicator of its health). Despite notable economic and social gains, the destruction of forests in order to develop land or create farms leads to widespread biodiversity losses and species endangerment in adjacent forests.

The Pulitzer Prize–winning scientist Edward Osborne Wilson helped develop the term biodiversity in the 1980s to recognize the collective value of ecosystems such as forests. Wilson's landmark 1988 book, *The Biodiversity of Life*, helped lay the groundwork for a new field of science that combined elements of biology, ecology, genetics, population dynamics, and species evolution. Wilson's work emphasized the fact that human resource consumption and population growth had led to a threat to the survival of innumerable species and thus affected the global genetic pool. In 1998, Wilson delivered a speech in the U.S. Senate on this issue, stating,

> Now when you cut a forest, an ancient forest in particular, you are not just removing a lot of big trees and a few birds fluttering around in the canopy. You are removing or drastically imperiling a vast array of species even within a few square miles of you. The number of these species may go to tens of thousands. Many of them, the very smallest of them, are still unknown to science, and science has not yet discovered the key role undoubtedly played in the maintenance of that ecosystem, as in the case of fungi, microorganisms, and many of the insects.
>
> They have been a symbol in these habitats, literally over millions of years of evolution. They are exquisitely adapted to these habitats. That is how the balance has been maintained.

Source: Speech given at the U.S. Senate on April 28, 1998, on-line at www.saveamericasforests.org/wilson/intro.htm.

Although the number of farms certainly will increase in the future, mainly in tropical regions, the United Nations is working to ensure that information on good agricultural practices is shared in developing countries. Measures include reducing runoff and capturing precipitation by preserving the stands of trees that separate and border fields; planting additional trees near fields so they can play a role in integrated pest management (some forest insects eat other insect species that harm farm crops); using forests strategically to limit erosion caused by wind and water; rotating crops and tilling the soil to let fields go fallow in order to support browsing forest species; and ceasing the release of animal waste, chemicals, fertilizers, and other toxic products into water supplies or adjacent forests.

These concepts and practices are being introduced in the Philippines. This Southeast Asian nation is made up of 7,100 different islands containing up to 115,831 square miles (300,002 square kilometers) of upland and 48,884 square miles (126,610 square kilometers) of tropical forests. Up to 18 percent of the country's economy relies on agriculture and forestry. The population of the Philippines is around 87 million, and resource consumption by this growing population has reduced forested areas by 60 percent.

In 2003, the government began programs to instruct farmers how to simultaneously protect the tropical forests and generate reliable agricultural harvests. Educational methods include seminars and demonstrations on good agricultural practices. Concepts that work well in the Philippines include intercropping trees and vegetables, planting tree farms, introducing new methods to grow crops on slopes, and using chemicals and fertilizers more efficiently.

After deforestation, erosion is the most pressing issue in the Philippines; 87,259 square miles (226,001 square kilometers) of cleared land has experienced severe erosion, impacting land stability, water quality, and general forest health. Efforts to educate local farmers will take years to complete, as the archipelago spans seventeen regions, 1,495 municipalities, and 41,956 towns.

Northern and Southern Hemisphere Forests

Human development has occurred at different paces around the world. Many northern nations, such as those in Europe and North America, went through industrial revolutions, massive economic expansions, and population surges between 1700 and 1900. Such industrialization resulted in unprecedented economic growth and, consequently, intensive resource consumption. Forests across the Northern Hemisphere were clear-cut (up to 85 percent in some places) to satisfy growing economic demands for wood products, building materials, fuel, and by-products.

Most southern locations in Pacific Asia, Africa, and South America did not reach a similar economic boom until about 1900. While the north gained economic wealth by consuming its forested regions 200 years earlier, the south did not commence a similar process until the beginning of the twentieth century. This dichotomy also resulted in a significant shift of resource consumption from the north to the south. While northern forests now are growing in size, as logging has been reduced and many farms have reverted to forests, logging and farming have become major focuses in the Southern Hemisphere.

Between 2000 and 2005, the northern temperate forests grew by 19,691 square miles (51,000 square kilometers) per year, while the southern tropical forested areas shrunk at an annual rate of 31,660 square miles (81,999 square kilometers). The countries with the greatest forest losses are Brazil, the Democratic Republic of the Congo, Indonesia, Myanmar, Nigeria, Sudan, Tanzania, Venezuela, and Zambia. The countries with greatest forest gain are Bulgaria, China, France, Italy, Portugal, Spain, and the United States.

Since the early 1990s, there have been considerable forest conservation efforts in such countries as Brazil, Venezuela, Indonesia, and Sudan. Some have been initiated by home nations seeking to better manage their natural resources. Many, however, have been the result of the efforts of the international community, including stewardship programming from the governmental, scientific, and conservation sectors.

These actions are based on the premise that many countries in the Southern Hemisphere have logged their forests too quickly, causing significant impacts. Such impacts include local climate alterations such as temperature increases, reduced rainfall and increased drought, increased erosion, and loss of biodiversity. And while logging produces profits in the short term, the long-term economic effects are not positive, resulting in little long-term financial stability for the region.

While the encouragement to protect southern forests has come predominantly from countries in the Northern Hemisphere, deforestation often has been due to the Northern Hemisphere's economic demand for tropical wood products. A number of partnership programs have been developed in response to this demand and consequent rapid forest loss.

The World Bank, with headquarters in Washington, D.C., often works to finance large infrastructure projects with developing countries. The World Bank now scrutinizes these financial agreements to limit tropical deforestation by implementing forest conservation and management efforts. Although historically criticized for poor oversight on loans for projects that caused environmental harm—such as building factories and roads, dredging shipping channels, and building dams—The World Bank has made measureable gains in terms of forest conservation and management issues in recent years.

In 2002, The World Bank released *Forests Strategy*, a planning document that sets policies to improve the conservation of forestland. As loan amounts and conditions increased over the years, a more detailed guidebook was needed to help government officials manage both resources and business growth. In 2008, The World Bank released *The Forests Sourcebook*, an updated resource for borrowing countries that helps to guide them through methods to best conserve forested areas while allowing sustainable economic development.

Deforestation and Climate Change

In addition to human-generated pollution such as vehicle and factory emissions, deforestation also has played a prominent role in global warming and climate change. According to the United Nations Food and Agriculture Organization, up to 20 percent of all carbon dioxide emitted annually, nearly 6 gigatons (6 billion tons), occurs in the form of forest removals.

The 1994 United Nations Framework Convention on Climate Change (UNFCCC), ratified by 192 countries, recognizes that climate and biomes are

---

THE FORESTS SOURCEBOOK, EXCERPT

Sustainable forest management is more than just growing and protecting trees—it is highly complex and can only be addressed through a range of actions that blend technical aspects of forestry with other considerations, such as how to strengthen policy and governance frameworks . . . Sustainably managing this resource for the vital global public environmental services it delivers will also require the participation of all stakeholders, from international donors to communities whose livelihoods depend on the forests.

—Warren Evans, The World Bank Director for Environment,
in *The Forests Sourcebook,* 2008.

interconnected and that human-produced pollution such as greenhouse gas emissions requires modification. The main goal of the convention is to set standards that reduce pollution to the point where the climate remains stable.

A 2005 proposal to the convention is a new program called Reducing Emissions from Deforestation and Land Degradation (REDD). This concept was first discussed in 2005 at the UNFCCC meeting in Montreal. REDD was introduced by Papua New Guinea and Costa Rica and is a system of monetary payments from industrial countries to developing ones to protect their forests for their carbon value.

Still in the early stages of development, REDD is intended to reduce tropical deforestation by directing funds to local governments to improve their capacity to manage forest areas with a number of different planning, research, enforcement, and other tools. A similar program was developed in Costa Rica at the end of the twentieth century. In these agreements, private land owners were paid to not log their land.

Public debate still is active on this new issue. The Center for International Forestry Research, based in Bali, Indonesia, is an organization working toward conserving forests and protecting its residents. In 2007, the center completed a multi-year study concluding that the REDD efforts will fail unless they properly address the greater underlying local political, economic, and social forces that result in forests losses. These forces include government corruption and policies that favor infrastructure development such as roads and bridges and short-term economic development gains that encourage logging, manufacturing, and converting forests to farmland.

Each country and each development project, argues the report, "require different solutions." An activity such as logging is not simply motivated by wood sales but often by the desire for economic development through building a new road or a factory complex, or by fulfilling agricultural needs. Regional problems may require regional solutions instead of a one-size-fits-all monetary payment scheme. Some experts in the forest management field feel that the controversy over REDD also may be about perspective: What outsiders (northern industrial countries) may view as an avoidable impact on forests may be something the local people (southern tropical communities) cannot do without.

For example, in western Africa, forests are commonly cut to provide fuel. Communities cannot survive without wood or charcoal to cook and heat their homes. In Central America, new roads are causing forest losses. Some of this infrastructure construction is to meet the transportaion needs of a growing, modern population. In South America, stronger economic development has resulted in more produce and cattle farms, as more products have been exported and global demand continues to grow. In Southeast Asia, the desire for palm oil drives many forest-to-farm conversions. The use of palm oil in fuel both supports the local economy and helps reduce local reliance on imported oil.

Although the issue of climate change has taken center stage in the international community, the measures taken to reduce deforestation represent a significant positive step for local communities. Simply ceasing to log any more forests is not a realistic option, which is why collaborative, reasonable solutions need to be developed and applied over time.

Selected Web Sites

Center for International Forestry Research: http://www.cifor.cgiar.org.
The Forests Dialogue: http://www.theforestsdialogue.org.
Landsat Program: http://landsat.gsfc.nasa.gov.
United Nations Forum on Forests: http://www.un.org/esa/forests/.
United Nations Framework Convention on Climate Change: http://unfccc.int/2860.php.
United Nations Good Agricultural Practices: http://www.fao.org/prods/GAP/home/principles_en.htm.

Further Reading

Allaby, Michael. *Temperate Forests.* New York: Chelsea House, 2006.
Faylon, Patricio. *Philippine Agriculture.* Los Baños, Philippines: Philippine Council for Agriculture, Forestry and Natural Resources Research and Development, 2007.
Gay, Kathlyn. *Rainforests of the World: A Reference Handbook.* Santa Barbara: ABC-CLIO, 2001.
Gibson, Clark, ed. *People and Forests.* Cambridge, MA: MIT Press, 2000.
Kanninen, Markku, et al. *Do Trees Grow on Money?* Bogor Barat, Indonesia: The Center for International Forestry Research, 2007.
Kramme, Linda, ed. *Practical Applications to Combat Illegal Logging.* New Haven, CT: Forest Dialogue, 2005.
Methot, Pierre. *Developing a Forest Transparency Initiative.* Washington, DC: World Resources Institute, 2006.
Wilson, Edward O. *The Diversity of Life.* 1988. New York: W.W. Norton, 2000.
Woodwell, George. *Forests in a Full World.* New Haven, CT: Yale University Press, 2001.
The World Bank. *The Forest Sourcebook.* Washington, DC: The World Bank, 2008

11 | Forest Stability

Forests provide considerable support for living organisms across the planet. Often defined by a high density of trees and plants, these ecological areas function as a habitat for diverse species all over the world. Even after millions of years of evolution on separate continents, some related species share multiple similarities due to the stability of the established forests they reside in.

For example, the white-tailed deer (*Odocoileus virginianus*) feeds on forest plants, including young tree saplings and acorns, in temperate forests across North America. Its distant cousin, the fallow deer (*dama*), lives in southern Europe; it is similar in size and also browses on young plants, tree bark, and apples in dense or open woodlands. Separated by the Atlantic Ocean more than 200 million years ago, these deer species could have evolved to be considerably different from each other, but instead both are medium-sized, live in family groups, and eat similar foods. A stable forest environment on two different continents has provided the right balance of food and shelter over time to allow both species to live and reproduce successfully.

Forests provide many positive benefits for both terrestrial and aquatic life. On land, forests reduce erosion, stabilize soils, and hold nutrients and carbon. When in contact with water, forests sustain thousands of tree, plant, and animal species and anchor large quantities of water for storage at the surface and underground. Forests provide dozens of types of habitat biomes that support millions of species, from insects to herbivores to wide-ranging birds. This web of life relies on the forest for food, water, and shelter.

Though a burnt forest may seem like a wasteland, it actually helps to nourish and attract new life. The old, burnt trees provide nutrients for young trees and also can attract bird species that find food (such as insects) and make their homes in them. The new shrubs and ground vegetation also attract new wildlife. *(Mike Goldwater/Getty Images)*

Forests also play a key role in atmospheric interchange. They generate massive amounts of oxygen and consume carbon dioxide. They create their own weather systems, contributing to cloud formation and encouraging condensation and, eventually, rain. Forests are host to microclimates where tree foliage absorbs the sun without heating up the understory. A dense stand of trees holds up to 75 percent of the moisture it receives through precipitation, creating a stable temperature and consistently humid environment for host species.

Forests have a unique ability to store large amounts of carbon in wood, in leaves, and in the soil. Through photosynthesis and growth, each mature deciduous tree absorbs 13 pounds (6 kilograms) of carbon per year and 1,950 pounds (885 kilograms) in an average 150-year life span. By banking the nutrients, the forest holds carbon for its future needs.

After a forest fire has occurred, the landscape may look like a wasteland, but fires are an important part of the forest life cycle. The ash from burned plants and trees contains high amounts of carbon that are deposited back into the soils. These nutrients help regrowth of trees in a new forest.

The natural death of a single tree in a forest is followed by its slow decomposition, a process that benefits birds (including woodpeckers), insects (especially termites), and various fungi. Eventually, new plant species emerge from the soil that has been enriched by the carbon released by the decomposing trees.

Forests at Risk

Forest densities have fluctuated since tree species appeared some 420 million years ago. Periods of warming temperatures resulted in excess carbon dioxide, oxygen vapor, and other greenhouse gases in the atmosphere and longer growing seasons, which produced more trees. At the other extreme, ice ages brought on much colder periods, during which widespread glaciers covering the land on several continents resulted in fewer trees.

Up to 4,000 years ago, 55 percent of the world was covered with forests, totaling at least 31 million square miles (80 million square kilometers). In 2008, there were about 13.8 million square miles (35.7 million square kilometers) of forests spread across all the continents except Antarctica.

Why such a vast reduction? The primary reason is dramatic human population growth, which has led to increased resource consumption, including logging for forest products and conversion to farms. The 140 years between 1850 and 1990 saw up to 965,000 square miles (2.5 million square kilometers) of boreal and temperate forests and 2.3 million square miles (6 million square kilometers) of tropical forest removed to create new farmland.

By the end of the twentieth century, up to 40,000 square miles (103,600 square kilometers) of tropical trees and 15,000 square miles (38,850 square kilometers) of temperate forests were harvested annually. Some areas have been more heavily logged than others.

In the Mediterranean region, including southern Europe, the Middle East, and North Africa, intensive logging has depleted the region to one-sixth of its original forests. Often, all that remains of these forests is in isolated, steep areas. In arid North Africa, only 1.2 percent of the land is forested. In the Middle East, only 1.9 percent is forested.

The United States was 46 percent forested in the 1650s, but, by 1992, the country had lost two-thirds of its forests due to widespread logging. The forest products industry cut 19 billion cubic feet (0.5 billion cubic meters) of wood in 1998, with half of it being used for construction of homes and other building, 30 percent used to make paper products, and 20 percent used for wood products such as furniture, cardboard and other paper products, and resins.

In 2008, the forest products industry in the United States, which has a strong international presence, sold $290 billion worth of products. It employed more than 1.4 million people.

Impacts on Primates

Although numerous forest ecosystems are closely regulated by home countries, the continuing fast pace of clear-cutting in regions such as the tropics is having a perilous impact on native animal species, especially primates—humans' closest relative.

There are approximately 394 different species of primates located in twenty-one countries across the Asian and African tropical forests. They include apes, lemurs, monkeys, and orangutans. The World Conservation Union estimated in 2007 that 114 of these species are threatened due to forest habitat loss and poaching.

The Malabar slender loris (*Loris tardigradus malabaricus*), about the size of a chipmunk, is considered an evolutionarily distinct primate species. It resides only in the tropical rain forests of Sri Lanka and southern India and forages during the night on a diet mainly of insects, but it also eats leaves, flowers, and plant shoots. This native species is endangered due to poaching by locals for the loris's perceived medicinal value and as a result of habitat destruction from forest logging. (David Haring/Photolibrary/Getty Images)

In particular, there are twenty-five primates that are close to extinction residing in the countries of Kenya (Tana River red colobus, *Procolobus rufomitratus*), Madagascar (silky sifaka, *Propithecus candidus*), Peru (Peruvian yellow-tailed woolly monkey, *Oreonax flavicauda*), Sri Lanka (Horton Plains slender loris, *Loris tardigradus nyctoceboides*), Vietnam (golden-headed langur, *Trachypithecus poliocephalus*), as well as in Colombia, Equatorial Guinea, Ghana, Indonesia, Tanzania, and Venezuela.

These primates rely on forested habitats that cover 2 percent of Earth's land surface but hold 50 percent of its living plants and animals. The majority of such threatened forested areas are in Asia, as this region underwent considerable logging growth over the last two decades of the twentieth century.

New Efforts to Conserve Forests

Recognizing the value of managing forests for both short- and long-term sustainability, many nations implemented regulatory measures in the last half century. These protective laws or policies allow limited logging. Two modern accounts, one in India and one in Australia, provide examples of approaches that take into consideration social, economic, and political conditions.

India

Located in South Asia, India occupies 1.2 million square miles (3.1 million square kilometers) and has a population of approximately 1.1 billion people. With the world's second-largest population, India has the twelfth-largest economy; the annual per capita income is around $4,100.

Stretching south to the Indian Ocean, north to the Himalayan Mountains, west to Pakistan, and east to China, this nation is 23 percent forested, with forestlands covering 276,000 square miles (714,840 square kilometers). Of this forest, up to 37 percent is tropical moist deciduous, 29 percent tropical dry, 8 percent tropical evergreen, and 26 percent temperate mixed deciduous and coniferous.

Most of the country's forests are concentrated in rural regions with poor soils. Up to 375 million Indians either live in or adjacent to forested areas, 100 million of whom rely on the forest environment for their daily needs.

The British government, which occupied much of the Indian subcontinent from 1858 to 1947, implemented a series of forest management practices in the nineteenth century. One effort designated certain royal crown lands to be used solely to build boats, for the benefit and strength of the British Empire.

In 1947, India attained its independence but continued to employ British forestry management practices. The Indian government regulated forest access but allowed considerable commercial logging. Over time, these forests were depleted, and between 1952 and 1980 up to 11,583 square miles (30,000 square kilometers) of tree plantations were established to increase the amount of raw materials available for the private wood industry. However, Indian forests were still being overlogged. As a result, up to 6 billion tons of topsoil were washed away from clear-cut land each year. This runoff also degraded water quality. The overall effect was a decrease in plant, fish, and other animal populations.

In 1988, a new law was implemented in response to widespread deforestation. While the new law still allowed logging to provide sufficient fuelwood, it declared that the commercial logging industry was not to be given priority, and long-term subsistence was to be the new emphasis for forest management. Environmental conservation, the policy stated, was paramount. The government also increased the size of forest reserves from 100,386 to 177,606 square miles (260,000 to 460,000 square kilometers).

By 1990, the Indian government had integrated public input into its forest planning and management process. Eighteen states, covering 80 percent of the forested regions, developed agreements with local villages to oversee forests and share in sales of timber. By 2007, this local control, coupled with conservation practices, had resulted in an increase in forested areas from 23 percent to nearly 33 percent. This twenty-year period of forest regrowth is a model studied around the world.

Australia

Bordered by the Indian and Pacific oceans in the Southern Hemisphere, Australia is the smallest of the seven continents. Australia has 2.9 million square miles (7.5 million square kilometers) of upland that is mostly arid terrain. In 2009, it contained approximately 22.1 million people, spread out at a low density of 6.8 people per square mile. The major population centers are located in the coastal regions.

Indigenous peoples called Aborigines settled the continent up to 42,000 years ago. Europeans arrived and began colonizing the island in 1788. These colonists found an estimated 3 million square miles (7.8 million square kilometers) of virgin forests on the island nation. Within a few decades, sections of the dense forests (266,410 square miles, or 690,002 square kilometers) and open forests (606,180 square miles, or 1.6 million square kilometers) had been selectively logged, with little governmental oversight.

By 1875, 45 percent of the original forests had been cutover, leaving only 1.3 million square miles (3.4 million square kilometers). In 1880, the Australian government created a forestry agency to manage public forests. While a steady supply of timber was a central goal, the government inventoried available wood and limited large clear-cuts.

Despite efforts to slow logging, Australia's forests continued to shrink in the twentieth century to 575,292 square miles (1.5 million square kilometers), mostly concentrated in the north and southeast. The wood chip industry saw the most growth, as it provided everything from paper products to fuel, mostly from conifer trees.

During the 1970s, Australia faced a number of high-profile logging protests, asking for better forest protection laws especially in southeastern coastal regions of the country. International companies, especially those from Japan, became the focus of protests due to their export of "homegrown" resources.

Public concern over deforestation led to a detailed catalog of natural resources and two key laws: the 1974 Environmental Protection Act and the 1975 National Parks and Wildlife Act. This new legislation, along with the government-drafted *National Conservation Strategy,* enacted conservation measures and applied pressure to forest products companies. In addition, a number of larger forest parcels were protected to preserve Australia's unique and mostly endemic biodiversity.

When the government refused to renew some wood chip export licenses, multiple legal challenges to these forest protection laws were mounted and upheld in the courts. In response, up to 335 square miles (868 square kilometers) of commercial tree plantations were created to provide softwood for a number of building, heating, and industrial products. In 2004, the forest products industry had shrunk somewhat but still employed 82,500 residents and earned $11 billion annually.

Australia's 2004 *State of the Forest* report shows gains in conservation. Twenty-six percent of Australian forests are protected from all logging. Up to 13 percent of the country's native forests are designated as bioreserves under federal law.

In 1988, up to 2,108 square miles (5,460 square kilometers) of Australia's forests were cut. This rate was reduced to 926 square miles (2,398 square kilometers) in 2002. However, 70 percent of the forests are owned and managed privately, and despite protective measures, the country still suffers from institutional problems stemming from weak enforcement and regulatory oversight. While the rate of loss has slowed, the nation still is battling to further reduce the loss of its forests as the demand for forest products and more farmland continues to grow.

FOREST ORGANIZATIONS

Name, Headquarters, and Web Site	Central Purpose
African Forest Research Network (AFORNET) Nairobi, Kenya http://www.afornet.org	Working with a network of research scientists, AFORENET promotes quality research on the use, management, and conservation of African forests.
Center for International Forestry Research (CIFOR) Bogor Barat, Indonesia http://www.cifor.cgiar.org	CIFOR uses high-impact research to conserve forests, while at the same time improving livelihoods, in worldwide tropical regions.
European Forest Institute Joensuu, Finland http://www.efi.int	Established by European nations, this institute performs scientific research on forests and policy issues.
International Union for Conservation of Nature (IUCN) Gland, Switzerland http://www.iucn.org	The oldest and largest global environmental network, the IUCN performs forest research and manages field projects around the world.
International Union of Forest Research Organizations (IUFRO) Vienna, Austria http://www.iufro.org	Linking more than 15,000 scientists across 110 countries, IUFRO coordinates global cooperation on forest research and its distribution.
Rainforest Alliance New York City http://www.rainforest-alliance.org	This alliance seeks to conserve biodiversity by transforming land-use practices in rain forests around the world.
United Nations Food and Agriculture Organization Washington, D.C. http://www.fao.org/forestry/en/	A central focus of this international agency is forestry, including conservation policies and sustainable development.
World Wildlife Fund (WWF) Washington, D.C. http://www.worldwildlife.org	For more than forty-five years, the WWF has been working with local communities to protect forest habitats in 100 countries.

Source: United Nations Forum on Forests, 2007.

International Collaboration

More humans rely on tree consumption for daily needs and economic viability than ever before. The global private free markets that helped develop a worldwide wood products industry, however, have failed to apply sufficient self-regulation and planning to sustaining this renewable resource.

Since the 1990s, a number of international and private forest conservation organizations have been formed in response to the rapid rate of deforestation. Some organizations work at the grassroots level, purchasing small parcels of land to protect them from development, for example. Others have an international presence and large staffs with scientific expertise. These conservation organizations study the health of and impacts on forests and have spent years inventorying forest habitats and species, educating the public, and enhancing efforts to better care for and protect forests.

Still other groups represent the logging and wood products manufacturers and make a case for their role in the economy. They point out that such manufacturers are responsible for creating thousands of jobs and that many are making efforts to use more ecologically sound practices to harvest and replant cutover areas.

The United Nations Conference on Environment and Development, also known as the Earth Summit, was held in 1992 in Rio de Janeiro, Brazil. It marked a significant milestone in international efforts to advance environmental protection. However, at the conference only a "non-binding statement of principles" was adopted with regard to forest conservation. This raised the ire of nations and organizations hoping for stronger regulations.

In 1995, a group of scientists, policy experts, and former government leaders moved to create an international body to strengthen work toward sustainable forest use. This group, named the World Commission on Forests and Sustainable Development (WCFSD), sought three key goals: to better link the Northern and Southern hemispheres with regard to policy issues; to further develop the science on forests to allow more thorough policy decisions; and to raise awareness of the dual role that forests play in protecting the natural environment and contributing to sustainable economies.

The WCFSD's 1999 report, *Our Forests, Our Future,* sent a direct message:

We are drawing on the world's natural capital far more rapidly than it is regenerating. Rather than living on the interest of the natural capital, we are borrowing from poorer communities and future generations.

This document outlined a crisis of global proportions that was the result of accelerated deforestation. To gather evidence, the WCFSD held five hearings across Asia, the Caribbean, Europe, Latin America, and North America. A common theme of the testimony from business members, citizens, government employees, and political leaders was that the current approach to managing forests was not working. They criticized systems that relied on subsidies and tax incentives with little intent to address corruption, weak regulatory frameworks, or the artificially low price of wood. *Our Forests, Our Future* has received widespread attention and is credited with playing a pivotal role in elevating the issue of forest overuse on the agenda of the international community.

Another example of international collaboration began in 1988, when The World Bank, a major international lender for infrastructure projects in developing countries, created a Working Group on Sustainable Forestry. This collective, made up of environmental and logging interests, discussed ways to improve forestry management. They drafted basic recommendations and guiding principles.

In 2000 this effort evolved into a second group called The Forests Dialogue. Also made up of both timber companies and forest protection organizations, including the American Forest Foundation, Conservation International, International Paper, and the World Business Council for Sustainable Development, The Forests Dialogue has held several separate meetings around the world to try and problem solve complex forestry issues.

In 2005, a meeting in Hong Kong focused on illegal logging. The 120 participants defined two central goals: to raise awareness in the business community of the impacts of illegal logging, and to identify solutions to improve forest management practices by the logging industry. Implementation of these concepts on a widespread basis is an ongoing challenge.

Following the groundwork laid on forest issues at the Earth Summit in 1992 and the contributions of the World Commission on Forests and Sustainable Development, the United Nations created a new organization called the Forum on the Forests in 2000 to address the "management, conservation, and sustainable development of all types of forest." The forum was charged with the task of implementing forest use agreements, enhancing cooperative work on forest conservation, and strengthening the political resolve of countries to accomplish forest protection goals.

The Forum on the Forests developed a landmark document that was approved by the full United Nations in 2007. This framework, called *The Agreement on All Types of Forests*, outlines goals to limit forest degradation, promote sustainable forest practices, reduce illegal deforestation, and respond to poverty in forested regions of the world.

One central premise of this document, is a strategy called the "Community Forests Approach," a central concept of which was motivating local residents to safeguard their own forest resources. Because the nearly two decades of debate and discussion since the Earth Summit in 1992 did not see overall conservation but instead a 3 percent reduction in worldwide forests, forum stakeholders concluded that the best way to protect and manage these resources was to make ownership and use at a local level a central theme. This view asserts that the people who depend on forests for food, fuel, and income are much more likely to embrace the concepts of sustainable management than multinational corporations seeking short-term gains.

Forests and the Future

Forests have played a central role in humankind's social and economic development, providing essential heat, shelter, and food, for tens of thousands of years. At the beginning of the twentieth century, scientists had just begun to explain the highly evolved chemical and biological interconnectedness of forests. Trees grew back, forest managers knew, and there appeared to be plenty of forests to cut as needed. What was not known 100 years ago was how much population growth would accelerate forest consumption.

Massive industrial development in the last 150 years resulted in a six-and-a-half-fold increase of the world's population by 2009, to 6.7 billion. This number continues to expand by 82.8 million people per year. Rapid population growth led to the consumption of most of the European and North American forests by the 1950s and spurred similar consumption in the Southern Hemisphere in the last three decades of the twentieth century. This pattern of growth and consumption has resulted in what economists today call a "full world" scenario: Such a large number of people means limited water, limited food sources, limited fuels, and, often, limited trees. Widespread clear-cuts have left only 13.8 million square miles (35.7 million square kilometers) of forest, covering 28 percent of the upland in 2008, and hosting a documented 1.75 million species (14 percent of all living creatures). In comparison, forests spanned 44 percent of upland in 1800.

By the beginning of the twenty-first century, twenty-five countries had leveled almost their entire forests, eighteen had cut 95 percent, and eleven had cut 90 percent. Many nations have realized the hard way that forest health is much more than just removing trees and leaving the forest to recover on its own. Replanting efforts have become a significant, if not compulsory, component of restoring scarred areas where clear-cutting is performed.

The acknowledgement of this situation by traditional Western economists has elevated its visibility. A 2006 British government report titled the *Stern Review on the Economics of Climate Change* concluded that forests are an essential component for economic and ecological stability, as well as critical for carbon storage. Authored by the English economist Lord Nicholas Stern, the report emphasizes how forest communities protect humans from the consequences of climate change and how losing the forests would bring a dual devastation to human economies and to forest biota.

Around the world, communities have begun to recognize their forests as dynamic, living biomes that support tremendous biodiversity and require intensive oversight. People are looking past the concept of forests as just being a resource for wood. As of 2007, 9 percent of global forests had been given some level of legal protection.

This coalescing of the international community marks a positive note for the future of forests. And increased knowledge of and application of forest science and management is making a difference in preserving and maintaining forests worldwide. However, this ecological biome still faces a turbulent existence for the foreseeable future given the forecast for economic and population growth.

Selected Web Sites

Australian Forest Inventory: http://www.daff.gov.au/brs/forest-veg/nfi.

Indian Government Ministry of Environment and Forests: http://envfor.nic.in.

Stern Review on the Economics of Climate Change: http://www.hm-treasury.gov.uk/sternreview_index.htm.

United Nations Food and Agriculture Organization, State of the World's Forests: http://www.fao.org/forestry/site/sofo/en/.

United Nation's Forum on Forests: http://www.un.org/esa/forests/index.html.

World Commission on Forests and Sustainable Development: http://www.iisd.org/wcfsd/.

Further Reading

Evans, Julian, ed. *The Forests Handbook*. 2 vols. Cornwall, UK: Blackwell Science, 2001.

Hanson, Arthur, ed. *Our Forests, Our Future*. Cambridge, UK: Cambridge University Press/World Commission on Forests and Sustainable Development, 1999.

Humphreys, David. *Logjam: Deforestation and the Crisis of Global Governance*. London, UK: Earthscan, 2006.

Luoma, Jon. *The Hidden Forest*. Corvallis: Oregon State University Press, 2006.

Newton, Adrian. *Forest Ecology and Conservation*. New York: Oxford University Press, 2007.

Stern, Nicholas. *Stern Review on the Economics of Climate Change*. London, UK: British HM Treasury/British Office of Climate Change, 2006.

Williams, Michael. *Deforesting the Earth*. Chicago: University of Chicago Press, 2002.

Acid rain. Rain or other precipitation that is unusually acidic. It is the result of emissions of carbon, nitrogen, and sulfur (typically from industry), which react with water to form acids. Acid rain causes harm to animals, plants, soils, and structures.

Acre. An area of land measuring 43,560 square feet (4,047 square kilometers). A square 1-acre plot measures 209 feet by 209 feet (63.7 meters by 63.7 meters); a circular acre has a radius of 117.75 feet (35.9 meters).

Agroforestry. A method of farming that integrates trees and shrubs with crops and livestock to create a sustainable land-use system.

Angiosperm. A flowering or fruit-bearing plant whose seeds grow inside a contained ovary (such as an apple, cherry, or pear).

Aquifer. An underground layer of permeable rock, gravel, soil, or sand from which groundwater can be extracted.

Atmosphere. The mixture of gases surrounding the Earth. By volume, it consists of about 78.08 percent nitrogen, 20.95 percent oxygen, 0.93 percent argon, 0.036 percent carbon dioxide, and trace amounts of other gases, including helium, hydrogen, methane, and ozone.

Bacteria. Simple, single-celled microorganisms that live in soil, water, organic matter, and the bodies of plants and animals. Some bacteria are pathogenic— that is, responsible for causing infectious diseases.

Biodiversity. Biological diversity, indicated by the variety of living organisms and ecosystems in which they live.

Biome. A regional ecosystem that is characterized by a particular climate and biological community.

Biosphere. The ecological system made up of all living organisms on Earth.

Board foot. A unit of wood measuring 144 cubic inches (365.76 cubic centimeters). A 1-inch (2.54-centimeter) by 12-inch (30.48-centimeter) shelving board that is 1 foot (30.48 centimeters) long is equal to 1 board foot. Board foot volume is determined by length (in feet) by width (in inches) by thickness (in inches), divided by 12.

Boreal. A forest area in the northern latitudes that is characterized by coniferous forests and long, very cold winters.

Canopy. The cover on the forest floor formed by overlapping tree branches.

Carbon dioxide (CO$_2$). A gas that is found naturally in the atmosphere at a concentration of 0.036 percent. It is fundamental to life, required for respiration and photosynthesis.

Carbon neutral. A state in which no carbon is released into the environment, achieved through controlling the emission and absorption of carbon.

Carbon sink. A natural or man-made reservoir or habitat (such as forests, wetlands, or oceans) that can store carbon for long periods of time.

Clear-cut. The removal of all of the trees in a large area of forest at one time, causing considerable environmental damage, including habitat destruction, the loss of species, soil erosion, a decline in water quality, and temperature increase.

Climate change. Global variation in the Earth's climate over time. A recent phenomenon (measured over the last two centuries), it is attributable largely to human atmospheric pollution. Climate change results in temperature increases and other shifts in weather patterns, severe droughts and floods, rising sea levels, and biotic changes (often causing the loss of species in an area).

Conifer. A plant or tree with needle-like leaves whose seeds are produced inside distinctive cones that grow on branches.

Conservation. The protection, preservation, management, or restoration of natural resources, plants, and wildlife.

Deciduous. A plant or tree that grows photosynthetic leaves that are shed at the end of a growing season or during periods of extreme drought.

Decomposition. The process by which organic matter is broken down by dynamic forces such as weather, water, and microbes.

Deforestation. The clearing of trees from forests by humans, either by logging or by burning.

Discharge. The release of untreated waste into waterways, often following excessive precipitation that exceeds the capacity of a sewer system or waste processing facility.

Ecosystem. The interaction of a biological community and its environment.

Endangered (and threatened) species. A species that is in danger of extinction because the total population is insufficient to reproduce enough offspring to ensure its survival. A threatened species exhibits declining or dangerously low population but still has enough members to maintain or increase its numbers.

Endemic. A plant or animal that is native to and occupies a limited geographic area.

Environment. External conditions that affect an organism or biological community.

Epiphyte. A rootless and nonparasitic plant that grows on another plant or tree for mechanical support. Epiphytes can produce their own nutrients from photosynthesis and absorb water from the air.

Erosion. The removal of materials from a location as a result of wind or water currents.

Eutrophication. Algae blooms in aquatic ecosystems that are caused by excessive concentrations of organic and inorganic nutrients, including garden and lawn fertilizers and sewage system discharges. These blooms reduce dissolved oxygen in the water when dead plant material decomposes, causing other organisms, such as fish and crabs, to die.

Forest. An upland area containing a density of trees and woody vegetation.

Forestry. The science and practice of managing and using forests and their associated resources, particularly trees, for human benefit.

Greenhouse effect. A warming of the Earth caused by the presence of heat-trapping gases in the atmosphere—primarily carbon dioxide, nitrous oxide, ozone, water vapor, and methane—mimicking the effect of a greenhouse.

Groundwater. Water below the Earth's surface.

Gymnosperm. A class of seed-bearing plants and trees that produce "naked" seeds not contained within an ovary; instead, the seeds grow inside cones (as in pine trees) or stalks (such as those of ginkgo).

Habitat. An area in which a particular plant or animal can live, grow, and reproduce naturally.

Hazardous waste. Industrial and household chemicals and other wastes that are highly toxic to humans and the environment.

Herbicides. Chemicals that are used to kill or inhibit the growth of plants. Selective herbicides kill a specific type of plant, while nonselective herbicides kill all plants.

Hydrologic cycle. The continuous movement of water as a result of evaporation, precipitation, and groundwater or surface water flows.

Hydrology. The science of the movement and properties of water, both above ground and underground.

Inorganic chemicals. Mineral-based compounds that do not contain carbon among their principal elements (such as acids, metals, minerals, and salts).

Insecticides. Chemicals used to kill insects that are harmful to plant and animal communities.

Multiple use. The management of land or forest for more than one purpose, such as wood production, preservation of water and air quality, wildlife conservation, aesthetic considerations, and recreation.

National forest (United States). Federal lands, designated by statute or presidential executive order and administered by the U.S. Forest Service, that are set aside for mixed use, including regulated resource extraction and recreation.

National park (United States). An area of natural beauty or historical interest that is administered by the U.S. National Park Service. Development and resource extraction in national parks are more limited than in national forests.

National wilderness area (United States). Public lands designated by the U.S. Congress to receive additional legal protection to preserve their natural state. The National Wilderness Protection System coordinates the activities of four federal agencies: the Bureau of Land Management, the U.S. Forest Service, the National Park Service, and the U.S. Fish and Wildlife Service.

Nitrates. A form of naturally occurring nitrogen that is found in fertilizers, human waste, and landfills and can cause eutrophication in wetlands and open water.

Non-point source. A source of pollution that comes from a broad area (such as car and truck emissions) rather than a single point of origin (known as a point source).

Nutrients. Chemicals that are necessary for organic growth and reproduction. For example, the primary plant nutrients are nitrogen, phosphorus, and potassium.

Organic chemicals. A large family of carbon-based chemicals, such as methane and vitamin B. In comparison, inorganic chemicals often are mineral-based, such as chloride and sodium.

pH. A measurement of the acidity or alkalinity of a solution, referring to the potential (p) of the hydrogen (H) ion concentration. The pH scale ranges from 0 to 14; acidic substances have lower pH values, while alkaline or basic substances have higher values.

Photosynthesis. A chemical process in which carbon dioxide and water are converted into organic compounds, primarily sugars, when the chlorophyll-containing tissues of plants are exposed to sunlight. Oxygen is released as a by-product. Photosynthesis occurs in algae, phytoplankton, plants, and some bacteria.

Point source. A source of pollution that originates from a single point, such as the discharge end of a pipe or a power plant.

Salvage cut. The harvesting of dead or damaged trees or trees that are in danger of being killed by fire, insects, disease, flooding, or other natural factors in order to save their economic value.

Savanna. A type of forest found in tropical or subtropical regions that is characterized by widely spaced trees and open grasslands.

Shrub layer. A forest layer that sits above the floor and below the understory. It is dominated by woody vegetation that can reach 15 feet (4.6 meters) in height.

Silviculture. The science and practice of establishing, tending, and reproducing forest stands of desired characteristics.

Square mile. An imperial or U.S. unit of measure that is equal to 640 acres (259 hectares).

Storm water runoff. The potent mix of dirt, nutrients, oils, and trash that is flushed from the land by rain, ending up in wetlands, lakes, rivers, streams, oceans, and groundwater.

Succession. Changes in the structure and composition of an ecological community. Succession is driven by the immigration of new species and the competitive struggle among species for nutrients and space.

Taxa. The classification of organisms in an ordered, hierarchical system that indicates their natural relationships. The taxa system consists of the following categories: kingdom, phylum, class, order, family, genus, and species.

Total maximum daily load (TMDL). The quantity of chemicals (such as nitrogen and phosphorus) that can be discharged into a waterway from all recognized sources (such as businesses, farms, septic systems, and street runoff), while still maintaining applicable water quality standards.

Transpiration. The loss of water in the form of vapor from plant surfaces, especially the leaves, roots, and stems. The rate of transpiration is increased by temperature, wind speed, and light intensity or decreased by humidity in the surrounding air.

Understory. An area of forest between the shrub layer and the canopy. This area often contains growing middle-age trees.

Virgin forest. An old-growth forest that has not been disrupted by logging efforts.

Water (H_2O). A tasteless, odorless liquid formed by two parts of hydrogen and one part of oxygen.

Watershed. A portion of land where all of the water runoff drains into the same body of water; also called a drainage basin.

Wetland. A low-lying area where high levels of saturation by water provide a habitat that is critical to plant and animal life.

For additional Web Sites on more specific topics, please see the Web Site listings in individual chapters.

Amazon Conservation Team: http://www.amazonteam.org.

Boreal Forest: http://www.borealforest.org.

Center for International Forestry Research: http://www.cifor.cgiar.org.

Congo Basin Forest Partnership: http://www.cbfp.org.

European Forest Institute: http://www.efi.int/portal/.

Forest Peoples Programme: http//www.forestpeoples.org.

The Forests Dialogue: http://environment.yale.edu/tfd/.

Global Forest Watch: http://www.globalforestwatch.org.

National Biodiversity Institute, Costa Rica: http://www.inbio.ac.cr/en/.

Russian Federal Forestry Agency: http://www.rosleshoz.gov.ru/english.

Sequoia and Kings Canyon National Parks: http://www.nps.gov/seki/.

Sequoia National Forest: http://www.fs.fed.us/r5/sequoia/.

Smithsonian National Zoological Park, Amazonia: http://nationalzoo.si.edu/Animals/Amazonia/.

Southern Black Forest Nature Park: http://www.naturparksuedschwarzwald.de/en/index_en.php.

Taiga Rescue Network: http://www.taigarescue.org.

U.S. Forest Service: http://www.fs.fed.us.

United Nations Food and Agriculture Organization, Forestry: http://www.fao.org/forestry/en/.

United Nations Forum on Forests: http://www.un.org/esa/forests/.

World Commission on Forests and Sustainable Development: http://www.iisd.org/wcfsd/.

World Wildlife Fund (U.S. site): http://www.worldwildlife.org.

WWF International (global site): http://www.panda.org.

Allaby, Michael. *Temperate Forests*. New York: Facts on File, 1999.

———. *Tropical Forests*. New York: Chelsea House, 2006.

Allen, William. *Green Phoenix: Restoring the Tropical Forests of Guanacaste, Costa Rica*. New York: Oxford University Press, 2001.

Baker, Christopher. *Costa Rica*. New York: DK, 2005.

Barter, James. *The Rivers of the World—The Congo*. Chicago: Lucent, 2003.

Bates, Henry Walter. *In the Heart of the Amazon*. 1863. New York: Penguin, 2007.

Bekker, Henk. *The Black Forest*. Walpole, MA: Hunter, 2007.

Bettinger, Pete. *Forest Management and Planning*. Burlington, MA: Academic Press, 2009.

Bierregaard, Richard, ed. *Lessons from Amazonia*. New Haven, CT: Yale University Press, 2001.

Blanc, J.J., et al. *African Elephant Status Report*. Gland, Switzerland: The World Conservation Union, 2003.

Butler, Rhett. *Borneo*. San Francisco: Mongabay, 2007.

Carson, Walter. *Tropical Forest Community Ecology*. Hoboken, NJ: Wiley-Blackwell, 2008.

Davis, Lawrence. *Forest Management*. Long Grove, IL: Waveland, 2005.

Day, Trevor. *Taiga*. New York: Chelsea House, 2006.

De Camino, Ronnie. *Costa Rica: Forest Strategy and the Evolution of Land Use*. Washington, DC: World Bank, 2000.

Evans, Julian, ed. *The Forests Handbook*. 2 vols. Cornwall, UK: Blackwell Science, 2001.

Evelyn, John. *Sylva: Or, a Discourse of Forest Trees*. 1662. Whitefish, MT: Kessinger, 2007.

Frankie, Gordon. *Biodiversity Conservation in Costa Rica*. Berkeley: University of California Press, 2004.

Garbutt, Nick. *Wild Borneo*. Cambridge, MA: MIT Press, 2006.

Gay, Kathlyn. *Rainforests of the World*. Santa Barbara, CA: ABC-CLIO, 2001.

Gibson, Clark, ed. *People and Forests*. Cambridge, MA: MIT Press, 2000.

Giles-Vernick, Tamara. *Cutting the Vines of the Past*. Charlottesville: University Press of Virginia, 2002.

Goulding, Michael. *Atlas of the Amazon*. Washington, DC: Smithsonian, 2003.

Hanson, Arthur, ed. *Our Forests, Our Future: Report of the World Commission on Forests and Sustainable Development.* Cambridge, UK: Cambridge University Press, 1999.

Hartesveldt, Richard. *The Giant Sequoia of the Sierra Nevada.* 1975. Washington, DC: U.S. Department of the Interior, National Park Service, 2005.

Hays, Forbes. *Taiga.* Washington, DC: American University, Trade Environment Database, 1998.

Heale, Jay. *Democratic Republic of the Congo.* Tarrytown, NY: Marshall Cavendish, 1999.

Honnay, Olivier, ed. *Forest Biodiversity.* Oxfordshire, UK: CABI, 2004.

Humphreys, David. *Logjam: Deforestation and the Crisis of Global Governance.* London, UK: Earthscan, 2006.

Hunter, Malcom. *Maintaining Biodiversity in Forest Ecosystems.* Cambridge, UK: Cambridge University Press, 1999.

Johansson, Philip. *The Forested Taiga: A Web of Life.* Berkeley Heights, NJ: Enslow, 2004.

Kasischke, Eric, ed. *Fire, Climate Changes, and Carbon Cycling in the Boreal Forest.* New York: Springer, 2000.

Kimmins, James. *Forest Ecology.* San Francisco: Benjamin Cummins, 2003.

Kolga, Margus. *The Red Book of the Peoples of the Russian Empire.* Tallinn, Estonia: NGO Red Book, 2001.

Kuusela, Kullervo. *Forest Resources in Europe.* London, UK: Cambridge University Press, 1995.

London, Mark. *The Last Forest: The Amazon in the Age of Globalization.* New York: Random House, 2007.

Luoma, Jon. *The Hidden Forest.* Corvallis: Oregon State University Press, 2006.

Marsh, George Perkins. *Man and Nature.* 1864. Seattle: University of Washington Press, 2003.

McEvoy, Thom. *Positive Impact Forestry.* Washington, DC: Island, 2004.

Mock, Greg, ed. *Human Pressure on the Brazilian Amazon Forests.* Washington, DC: World Resources Institute, 2006.

Moeliono, Moe, ed. *The Decentralization of Forest Governance.* London, UK: Earthscan, 2008.

Montagnini, Florencia. *Tropical Forest Ecology.* New York: Springer, 2005.

Muruthi, Philip. *African Heartlands.* Washington, DC: Island, 2004.

Nelson, Andrew. *Protected Area Effectiveness in Reducing Tropical Deforestation.* Washington, DC: The World Bank, 2009.

Newton, Adrian. *Forest Ecology and Conservation.* New York: Oxford University Press, 2007.

Page, David. *Yosemite and the Southern Sierra Nevada.* Woodstock, VT: Countryman, 2008.

Pearce, David. *The Value of Forest Ecosystems.* Gland, Switzerland: Convention on Biological Diversity, 2001.

Perez, M. Ruiz. *Logging in the Congo Basin.* Amsterdam, The Netherlands: Elsevier, 2005.

Perry, David. *Forest Ecosystems.* Baltimore: Johns Hopkins University Press, 2008.

Philipp, Dorthee. *Schwarzwald: Black Forest.* Berlin, Germany: Art Stock, 2007.

Preston, Richard. *The Wild Trees.* New York: Random House, 2008.

Prunier, Gerard. *Africa's World War.* New York: Oxford University Press, 2008.

Raffaele, Paul. "Out of Time," *Smithsonian,* April 2005.

Robinson, George. *Sequoia and King's Canyon.* Watertown, MA: Sierra, 2006.

Russon, Anne. *Orangutans: Wizards of the Rain Forest.* Buffalo, NY: Firefly, 2004.

Scherer-Lorenzen, Michael, ed. *Forest Diversity and Function: Temperate and Boreal Systems.* New York: Springer, 2005.

Schilthuizen, Menno. *Borneo's Botanical Secret.* Jakarta, Indonesia: World Wildlife Fund, 2006.

Shugart, Herman, ed. *A Systems Analysis of the Global Boreal Forest.* New York: Cambridge University Press, 2005.

Starr, Chris. *Woodland Management.* Wiltshire, UK: Crowood, 2005.

Stern, Nicholas. *Stern Review on the Economics of Climate Change.* London, UK: British HM Treasury/British Office of Climate Change, 2006.

Tayler, Jeffery. *Facing the Congo.* St. Paul, MN: Ruminator, 2000.

U.S. Forest Service. *Giant Sequoia National Monument Management Plan.* Washington, DC: U.S. Department of Agriculture, U.S. Forest Service, 2003.

Vermaas, Lori. *Sequoia.* Washington, DC: Smithsonian, 2003.

Wadley, Reed, ed. *Histories of the Borneo Environment.* Leiden, The Netherlands: Kitlv, 2006.

Wadsworth, Ginger. *Giant Sequoia Trees.* Minneapolis, MN: Lerner, 1995.

Wallace, Scott. "Last of the Amazon," *National Geographic Magazine,* January 2007.

White, Mel. "Borneo's Moment of Truth," *National Geographic Magazine,* November 2008.

Williams, Michael. *Deforesting the Earth.* Chicago: University of Chicago Press, 2002.

Woodwell, George. *Forests in a Full World.* New Haven, CT: Yale University Press, 2001.

The World Bank. *The Forests Sourcebook.* Washington, DC: The World Bank, 2008.

World Conservation Union. *Gunung Mulu National Park: Borneo, Malaysia.* Gland, Switzerland: World Conservation Union, 2000.

Young, Raymond. *Introduction to Forest Ecosystem Science and Management.* Hoboken, NJ: John Wiley and Sons, 2003.

Index

Italic page numbers indicate images and figures.